PLANTS OF THE

Tahoe Basin

PLANTS OF THE

Tahoe Basin

FLOWERING PLANTS, TREES, AND FERNS

A   Photographic   Guide

MICHAEL   GRAF

CALIFORNIA NATIVE PLANT SOCIETY PRESS

UNIVERSITY OF CALIFORNIA PRESS
BERKELEY   LOS ANGELES   LONDON

Dedication

This book is dedicated to my parents, Peter and Susan Graf,
who instilled in me my love of nature and the outdoors, and who offered
me encouragement on this project every step of the way.

California Native Plant Society Press
Sacramento, California

University of California Press
Berkeley and Los Angeles, California

University of California Press, Ltd.
London, England

 1722 J Street, Suite 17
 Sacramento, CA 95814
 916-447-2677
 Email: cnps@cnps.org

Library of Congress Cataloging-in-Publication Data

Graf, Michael, 1961–
 Plants of the Tahoe Basin: Flowering Plants, Trees, and Ferns : a photographic
guide / Michael Graf.
  · p. cm.
 Includes bibliographical references.
 ISBN: 0-520-21583-4 (cloth : alk. paper). — ISBN: 0-520-21541-9 (pbk. :
alk. paper)
  1. Botany—Tahoe, Lake, Region (Calif. and Nev.) 2. Botany—Tahoe, Lake,
Region (Calif. and Nev.) Pictorial works. 3. Plants Identification. I. Title.
QK149.G735 1999
581.9794'38 dc21   99-13267
                             CIP

First Edition

Photographs © 1999 Michael Graf

Edited and produced by Phyllis M. Faber, CNPS Press

Cover and book design and typesetting by Beth Hansen-Winter

Drawings courtesy of the Jepson Herbarium, University of California, Berkeley

Printed in Hong Kong through Global Interprint, Santa Rosa, CA

9 8 7 6 5 4 3 2 1

ACKNOWLEDGMENTS

Many people contributed to the writing of this book. I would like to thank the Jepson Herbarium and the California Academy of Sciences staff for allowing me free access to their marvelous collections. The researchers at Jepson, including Barbara Ertter, Lincoln Constance, and Bruce Baldwin, were extremely generous in sharing their vast knowledge. I would especially like to thank John Strother and Margriet Wetherwax for believing in this project from the outset, for being patient with my numerous questions on the sunflower and figwort families, and for reviewing early drafts. Many of *The Jepson Manual* authors, including Dean Taylor, John Mooring, James Morefield, David Keil, Dale McNeal, Tim Messick, Theresa Scholars, and Peter Hoch also graciously offered their expertise. Ted Barkley and Robert Price were particularly generous, helping me repeatedly on taxonomic issues in the sunflower and mustard families.

Many others helped with reviewing early manuscript drafts and/or offering their own expertise regarding Tahoe flora. These include Stephan Edwards, Arnold Tiehm, Wilma Follette, Toni Fauver, Amy Merrill, and Jamie Buffington. I am also grateful to the members of the Tahoe Chapter of CNPS, including Paul Cushing and especially Steve Matson, for showing me the great wildflower spots of the Basin, and for sharing their considerable local knowledge. I would like to thank Professor Laird Blackwell for allowing me to participate in his annual week-long wildflower course.

I would also like to thank Ann Hayes for editorial work on the introductory material, Nora Harlow for several rounds of proof-reading the manuscript, Beth Hansen-Winter of CNPS Press for applying her considerable design talents to create the product you hold in your hand, and Doris Kretschmer at UC

Press for educating me on a number of aspects of the publishing business, always in a positive and supportive manner.

Several individuals deserve special mention here. Phyllis Faber encouraged this book when it was nothing more than a rumor and a box of disorganized slides. Besides providing his own valuable input, Robert Ornduff expressed his faith in the early stages that this was a worthwhile project that could contribute to an appreciation of Tahoe's flora. Julie Carville's book *Lingering in Tahoe's Wild Gardens* was both instructive and tremendously inspirational to me as I set out into the field to learn the plants for the first time. Finally, I would reserve a special debt of gratitude for Gladys Smith, whose *A Flora of the Tahoe Basin and Neighboring Areas* was an invaluable resource as I struggled to learn Tahoe's many plant species. While subsequent research has revised some of Mrs. Smith's collection references, her flora remains an extraordinary source of information regarding the scope of Tahoe's floral diversity, as recorded by the botanical collections made in Tahoe over the last century. In offering this thanks, I am, of course, also owing a great deal to the many botanists who have traveled Tahoe's wild country before me. It is in that vein that I would wish this book to be considered, not in any way as a final product, but instead, as merely another chapter of a continuing work in progress. Happy hiking.

Contents

White fir trees loom above Emerald Bay with Lake Tahoe in the background.

INTRODUCTION

This book is a photographic guide to the flora of the Tahoe Basin. Photographic guidebooks that cover the entire Sierra Nevada often leave the wildflower enthusiast in the field yearning for more detail, particularly in an area with as much diversity as the Tahoe Basin. This book fills in some of that detail, providing along the way some basic information on the ecology of Tahoe's flora.

The text includes only the higher plants, which are characterized by vascular conducting tissue and, typically, well developed root and shoot systems. Tahoe's vascular plants may be generally divided into three main categories: ferns and their allies, conifers, and flowering plants. Although abundant in Tahoe, grasses, rushes, and sedges (which are flowering plants) are not covered here because of the complexity of their keying characteristics and identification. Also not covered are several other mostly aquatic plant families. (A list of plant families occurring in Tahoe but not covered in this book is at the end of the glossary.) Finally, this book is generally confined to the native plants of the area, but includes introduced species that have become naturalized in the Basin.

The geographic range of this book covers the entire Tahoe Basin, from the Desolation Wilderness on the west to the Carson Range on the east. For practical purposes, the Basin is defined to include Donner Lake and Donner Pass, Sagehen Meadows, Castle Peak, Pole Creek, Shirley Canyon, Granite Chief, and Alpine Meadows in the north and Hope Valley and the Carson Pass and Luther Pass areas in the south. The inclusion of these northerly and southerly locations is defensible in part on ecological grounds, for each represents a continuous extension of Tahoe Basin plant communities. More to the point, these areas offer spectacular wildflower viewing and thus should be included in any guide to Tahoe's flora.

ORGANIZATION

The photographed plants in this book are arranged alpha-
betically according to scientific classification, first by family and
then by genus and species. Other described species for which
no photograph is provided are presented in general, though
not always exact, alphabetical order. To find the description of
a specific plant, the reader should consult the index.

For the beginner, taxonomic organization may initially appear less useful than an arrangement based on more easily
observable criteria such as flower color. In the long run, however, taxonomic classification teaches the reader to use plant
characteristics, such as the four petals of the evening primrose family or the square stem of the mint family, to identify
plant taxa. Such characters generally will be applicable to all
family members, including species found outside the Tahoe
Basin. Taxonomic organization also offers insight into the evolutionary relationships of Tahoe's flora, leading the reader to
a fuller appreciation of the remarkable variation within the
plant kingdom.

Each description gives a plant's common and scientific
names. Scientific names consist of a plant's genus name, which
is always capitalized, and species name, which is in lower case.
In taxonomic hierarchy, each plant family has at least one genus (or genera in the plural), and each genus has one or more
species. For example, *Pinus jeffreyi* (Jeffrey pine) and *Pinus
monticola* (western white pine) are two species that belong to
the genus *Pinus*, which includes only species of pines, and to
the Pinaceae or pine family, which also includes common Basin conifers such as firs and hemlock.

Both common and scientific names are used in referring
to plants. While scientific names are sometimes difficult to
learn, they have the advantage of being universally accepted
as the exclusive identification for individual plant species. In
contrast, a plant species may have several common names

(which can vary by region) or lack a common name altogether. Taxonomic references in this book are according to the 1993 *Jepson Manual*, which at this time is the most authoritative work on the flora of California and is generally applicable to all plants found at Tahoe, including those occurring on the Nevada side.

For quick reference, each plant description begins with notes on the size, blooming period, and preferred habitat of the featured species. In conjunction with the color photo, this information should aid in confirming a plant's identification. The text provides additional information on the species, including plant characteristics, ecology, and historical uses. Closely related or similar-looking species are also described. More than 600 plant species found in the Basin are included.

How to Use this Book

For the enthusiastic hiker with little or no botanical knowledge, the many color photos offer the simplest means of plant identification. For the interested amateur, *Plants of the Tahoe Basin* also sets forth basic information on traits shared within families, genera, and species of Tahoe plants. Such common characteristics are extremely helpful in recognizing plants in the field.

The plant descriptions identify further characteristics that help to differentiate among genera and species. Whenever possible, the descriptions emphasize easily observable characteristics, such as the number of needles or flower petals. More detailed information on less obvious traits, such as leaf shape or the presence or absence of hairs, is provided to distinguish individual species. For some diverse genera in the Basin such as *Penstemon* or *Lupinus*, the use of a simple 10-16x hand lens can help clarify uncertainties regarding a specimen's identity. To further aid the reader, drawings of plant anatomy for the various plant groups covered in this book and a glossary of botanical terms are found at the end of this book.

Ultimately, knowledge of taxonomic characteristics is the easiest way to identify plants encountered in the field. As one learns the characters shared by various plant taxa, it becomes easier to determine the identity of a given specimen. Among flowering plants, floral structure is generally the most important factor in plant identification. Irregular flowers, for example, in which the petals (and sometimes sepals) are asymmetric, occur in a limited number of plant families in the Basin, notably the mints (Lamiaceae), peas (Fabaceae), figworts (Scrophulariaceae), orchids (Orchidaceae), and violets (Violaceae), as well as many species in the poppy (Papaveraceae) and buttercup (Ranunculaceae) families. The multi-flowered heads of the large sunflower family are easy to identify. The number of flower parts (petals or sepals) can also help to narrow the field. Only a few families in the Basin, for example, are characterized by four-petaled flowers, notably the mustards (Brassicaceae), the evening primroses (Onagraceae), the poppies, and the madders (Rubiaceae). Flower parts in threes or sixes make it extremely likely that a given plant is a member of the lily (Liliaceae), iris (Iridaceae), or orchid families.

Other floral characteristics can be equally important in identification. *Stamens* may extend past the corolla margins, or may remain below. *Anthers* and *stigmas* come in a variety of shapes, ranging from narrow clubs to four-lobed crosses. A useful characteristic in distinguishing flowering plant families is the position of the *ovary* relative to the attachment of the sepals and petals. Over evolutionary time, the sepals, petals, and stamens of flowers in many families have fused together to envelop the ovary. In such a case, the ovary is *inferior* since it appears to rest below the base of the remaining unfused parts. In flowers where such fusion has not occurred, the ovary rests above the base of the flower parts and is said to be *superior*. A good way to determine the location of the ovary is by observing a flower as it goes to seed. In species with inferior ovaries (such as in the evening primrose family), the fruit grows below the fading flower parts, while in species with superior ovaries (such as in the mustard family), the fruits grow above.

The shape, texture, and arrangement of leaves on a plant are also helpful keying tools. Leaves may be smooth-edged, serrated, lobed, compound, or needle-like. Their texture may be soft, hairy, waxy, or succulent. They may grow opposite from one another, alternately, or in whorls along the stem. Many plants have only *basal* leaves, which grow from the stem base. *Pinnately compound* leaves are uncommon in Tahoe, so where they appear they can be used to identify several species such as mountain ash (*Sorbus californica*) or members of the genus *Polemonium* in the phlox family. Most wildflower enthusiasts have learned to recognize the *palmately compound* leaves of the lupines (*Lupinus* spp.), one of the largest genera in the Basin.

Variations among plant characteristics are the result of plants' adaptations to the varied habitats and elevation ranges within the Tahoe Basin. Hikers may notice that plants in moist areas often have large, relatively thin leaves, while the leaves of plants growing in drier conditions typically will be small, thick, and either waxy or hairy, traits that provide basic protection against water loss through evapotranspiration. In windier conditions and variable levels of snowpack, high-elevation plants are generally shorter than their lower-elevation relatives, a comparison that can be readily observed among populations of species with wide elevation ranges such as lodgepole pine (*Pinus contorta* ssp. *murrayana*), woolly sunflower (*Eriophyllum lanatum*), Sierra wallflower (*Erysimum capitatum*), or showy penstemon (*Penstemon speciosus*). Among flowering plants, floral characteristics are also a direct result of natural selection as plants maximize reproductive success by furthering mutual relationships with animal pollinators.

Where possible, this book illustrates the relationship between a plant's features and its physical and biotic environment. To further the reader's knowledge of the relationship between Tahoe's environment and the resident flora, the following sections discuss the origins of the Tahoe flora and the vegetation ecology and resulting vegetation communities of the Basin. These sections are provided as general background to supplement the ecological material provided in the plant descriptions. Before turning to these topics, a brief discussion of plant taxonomy is set forth below.

TAXONOMIC ORGANIZATION OF THE PLANT KINGDOM

Most modern classification systems separate living organisms into either five or six kingdoms, the bacteria (one or two kingdoms), the protistans, the fungi, the animals, and the plants. The plant kingdom is generally organized into three groups comprising the nonvascular mosses, hornworts, and liverworts; the vascular non-seed plants, which include ferns and their allies; and the seed-bearing plants (which are also vascular). Seed-bearing plants are further separated into two subdivisions, those that bear their seeds "naked" on the plant's reproductive parts (the *gymnosperms*) and those that bear their seeds encased in a protective ovary (the *angiosperms*). The angiosperms, or flowering plants, are composed of two classes, the *monocotyledons* and the *dicotyledons*. At this time, taxonomists are not in complete agreement as to the placement of the uni- or multi-cellular photosynthetic organisms collectively known as algae. Some place algae in the kingdom Protista, based on their simpler reproductive mechanisms. Others (including *The Jepson Manual*) retain algae in the plant kingdom, because of their photosynthetic properties and ancestral lineage to the other three plant groups.

Vascular plants are generally defined as plants possessing vessel tissues that transport water and nutrient or organic solutes throughout the plant. The vascular non-seed plants are well represented in the Basin by ferns, club mosses, and horsetails. In the temperate northern hemisphere most gymnosperms are evergreen conifers, rep-

The lichen on this white fir log is formed by the union of an algae and a fungus, representing two different taxonomic kingdoms.

resented in the Tahoe Basin by the various species of pine, fir, hemlock, juniper, and incense-cedar. Angiosperms form the remainder of the Tahoe flora and comprise the vast majority of plant species in the Basin, from tiny two-inch annuals to the 100-foot-tall cottonwoods found in lower-elevation stream corridors.

Taxonomic organization of the plant kingdom is based upon evolutionary relationships, which are being continually redefined as knowledge is gained. Even when these relationships are well understood, decisions regarding where to draw taxonomic divisions are often difficult to make. Most botanists adopt a general rule that recognizes as separate species the smallest observable plant groups that are consistently distinct from one another over time. Where differences in plant morphology do not meet this definition, taxonomists may classify different populations as subspecies, varieties, or forms. Inevitably, a certain amount of arbitrariness creeps into this process. Most researchers agree, for example, that some of the large, inclusive families listed in *The Jepson Manual*, such as the heaths (Ericaceae) or the lilies (Liliaceae), could be split into smaller families without sacrificing taxonomic integrity. Readers should not be alarmed to encounter taxonomic treatments of plants in the Tahoe flora that differ from the one presented in this book (and in *The Jepson Manual*). These varying interpretations are little more than scientific disagreements as to how one imposes order on a natural system that appears at times unwilling to be so easily categorized.

Evolutionary Basis for Taxonomic Structure

Scientists confront a more fundamental taxonomic issue when questions arise regarding the evolutionary heritage of a plant species. A major challenge for botanists is to establish the direction of evolutionary changes, particularly among groups of plants that may have developed along entirely separate paths. For example, researchers for many years believed that *catkin*-type flowers, typical of the willow, beech, or birch families, were primitive because of their small size, lack of petals and sepals, and use of wind pollination, all traits similar to the ancestral gymnosperms. Further study revealed, however, that these flowers were relatively advanced, having *reduced* their floral parts in response to the reproductive opportunities available through wind pollination in their preferred habitats.

Such insights are gained by applying the principle that long-term evolutionary trends must be explainable by *natural selection.* Natural selection theory states that a character change will spread over time through a population only if the change confers some sort of survival advantage over other plants not possessing the new trait. Because flower characteristics play such a central role in the reproductive success of angiosperms, botanists have paid particular attention to flower structure in mapping evolutionary and taxonomic relationships.

The evolutionary development of flower structure can be characterized by two trends: simplification and differentiation. Simplification eliminates redundancy in function. In angiosperms, the number of flower parts, most notably pistils and stamens, has been reduced or, in the case of single-sex flowers, eliminated outright. Simplification can also cause flower parts performing similar functions to fuse together into single structures. Good examples of this phenomenon can be seen in *Penstemon* or in various genera of the phlox family, such as *Ipomopsis, Collomia, Gilia,* and *Phlox*, in which petals, and sometimes sepals, have fused together to form a continuous floral tube, often in response to selective pressures to reserve nectar for the most desirable pollinators.

Differentiation occurs when flower parts that perform separate functions, such as the anther and filament of the stamen, separate into distinct structures, each performing its respective job to further successful reproduction. Differentiation was responsible for the evolution of primitive petals and sepals from leaves and stamens, a transition that marked the beginning of angiosperm history. Differentiation may also occur in the same flower parts, as evidenced by the elaborately irregular flowers of the pea, mint, orchid, and figwort families, in which once identical petals have adopted separate forms that attract and host their preferred pollinators.

In the Tahoe Basin, one can observe primitive flower structure in Brown's peony (*Paeonia brownii*) and the various buttercups (*Ranunculus*). In the peony, one often sees leaves attached to the large outer sepals surrounding the petals, indicating incomplete differentiation between leaves and flower parts. Peonies also share with buttercups traits considered characteristic of the earliest flowers: unfused, often spirally arranged petals, numerous stamens and pistils, and a generally undiscriminating pollination strategy in which the fertile flower is visited by a number of different would-be pollinators.

Increasingly, evolutionary relationships among taxonomic groups are determined by molecular analysis of plant DNA molecules. Since these molecules bear the imprints of each species' accumulation of genetic mutations, DNA coding is the most reliable method of deducing evolutionary relationships. Because it is expensive and time-consuming, however, molecular research has been conducted only on a small number of species. Where it has been developed, DNA evidence in most instances corroborates the traditional evolutionary hierarchy based on plant characters and fossil evidence. In some cases, however, molecular research has led to the overhaul of long-standing taxonomic relationships. As more DNA work is done, additional changes can be expected.

ORIGINS OF TAHOE'S FLORA

The forerunners of the plants found in the Basin today probably were multicellular green algae that photosynthesized near the surface of shallow oceans over 700 million years ago. Approximately 200 million years later, plants made the transition to land by developing water-conserving cell walls known as *cuticles* that protected their vegetative and reproductive parts from drying out in the harsh new environment. Competition for resources quickly favored plants that developed vascular tissue, which provides structural support for above-ground growth and transports water, nutrients, and organic compounds throughout the plant. In developing vascular parts, plants exponentially increased their access to two elements necessary for photosynthesis, sunlight and carbon dioxide. This led to the rapid expansion, approximately 400 million years ago, of the seedless vascular plants, which thrived in the warm, humid conditions prevalent at that time. Today, these non-seed plants (which include the ferns and horsetails) continue to rely on moist conditions for spore germination and sexual reproduction.

About fifty million years after the development of the first vascular plants, the first primitive seed plants, forerunners of

the gymnosperms, appeared. The advent of drier, cooler inland climates as a result of continental drift and consolidation
created optimum conditions for the spread of gymnosperms
over the next 150 million years, a development that paralleled
the rise of the dinosaurs. Indeed, when we picture a scene from
the Jurassic period, we might imagine dinosaurs, in some regions at least, moving amidst the ancestors of our modern-
day redwoods, pines, and firs.

The seed plants were better suited to drier conditions for
several reasons. The development of the *pollen grain* meant
that mature sperm cells could travel by wind, protected by a
tough cellular wall. This freed the seed plants from dependence
on free water for fertilization and thus greatly enhanced the
ability of plants to cross-pollinate. The seed provided nour-
ishment for the young seedling and a hardened coat that would
not open until sufficient moisture was available, allowing for
improved dispersal and germination success. These evolutionary advancements enabled the gymnosperms to flourish in
areas in which surface moisture was not always available. Look-
ing out across Tahoe's conifer forests, we can observe the success of these advancements first-hand.

The ancestors of the first flowering plants probably lived in
tropical forests and relied, in part, on insects attracted to their
high-protein pollen grains to achieve cross-fertilization. Taking
advantage of this natural affinity, these plants developed more
specialized reproductive organs and flattened petal appendages
that attract pollinators (petals are thought to have derived from
both leaves and stamens) and gradually enclosed their developing seeds in a protective covering known as the *carpel*. Parts of
the flower still accessible to pollinators were often strongly
scented or filled with a sugar-rich fluid we call nectar.

The evolutionary development of the flower approximately
130 million years ago, and its subsequent coevolution with
insect and other animal pollinators, increased the frequency

of genetic mixing, allowing flowering plants to quickly diversify and fill a broad range of expanding environmental niches. This radiation was furthered by additional angiosperm developments, including a more efficient fluid transport system and a unique fertilization process, which results in nutrition-rich seeds that provide the angiosperm seedling needed energy during the vulnerable germination period before photosynthetic leaves are fully developed.

The red berries of mountain ash are especially visible to birds, which disperse the undigested seeds.

Another angiosperm innovation is its post-fertilization process, in which the female reproductive parts develop into a protective covering around the seed, which we call a *fruit*. The fruit offers greater protection to the enclosed seed and may also function as a means of dispersal in a number of ways. Fruits in many sunflower genera (as well as *Epilobium* or *Salix*) ride the air currents on long feathery appendages, finding new areas to colonize. Stickseed fruits (*Hackelia*) are burred, adhering to any textured object that rubs against them. Perhaps best known are the many edible fruits, well represented in the Basin by such families as gooseberry (Grossulariaceae), honeysuckle (Caprifoliaceae), and rose (Rosaceae), which turn bright colors at maturity and entice hungry animals, which disperse the undigested seeds, accompanied by a generous supply of fertilizer.

The diversification of angiosperms over the last 100 million years is one of the remarkable stories of evolution. In a relatively short time period, angiosperms spread into seasonally cold or dry temperate environments by developing thick cuticles on above-ground plant parts that slow water loss, or by developing deciduous leaves that grow on an annual basis with the onset of warmer temperatures and sufficient moisture. In areas where soils or precipitation levels did not support woody growth, angiosperms utilized the herbaceous growth form, in which the above-ground parts (or in the case of annuals, the entire plant) dies each year at the end of the growing season. In the Basin, the wide range of habitats occupied by the angiosperms, from swampy meadows to shady forests to wind-blown alpine summits, is a testament to the overall success of these diverse adaptations.

Geologic and Climatic History of the Tahoe Basin

The physical changes in the environment that fueled the radiation of flowering plants are well illustrated by the geologic and climatic history of the Tahoe Basin. This history is part of a larger story that has shaped the modern landscape and climate of California. According to the theory of *plate tectonics*, the upper surface of the earth's crust is covered by large plates that move slowly above the more fluid inner mantle. Plate movement accounts for the phenomenon of *continental drift*, in which continents move together to form large supercontinents and then break apart over millions of years. California's story began about 230 million years ago, when the North American Plate broke away from the Eurasian land mass and started moving west, opening a gap between the continents that was eventually filled by the Atlantic Ocean. In its westward migration the North American Plate bumped up against the Pacific Plate, forcing the denser oceanic plate to dive below in a process geologists refer to as *subduction*. Forced into the hot interior mantle, portions of the Pacific Plate melted, resulting in molten rock that rose and cooled slowly under the surface for millions of years, forming over time the massive granitic block known as the Sierra Batholith, which today comprises the underlying structure of the Sierra Nevada. The tremendous pressure and heat caused by the intrusion of the molten batholith metamorphosed adjacent subsurface deposits, exposed outcrops of which can be seen today in areas such as Mount Tallac.

The subduction of the Pacific Plate caused uplift of mountain ranges, including the early Sierra Nevada, over the next 200 million years. Natural forces continued to erode these ancient ranges, stripping away most of the overlying layers, exposing the underlying granite, and depositing large amounts of sediment into the Central Valley drainage basin. About twenty-five million years ago, at the beginning of the Miocene epoch, the Pacific and North American plates began to slide past each other instead of colliding headlong. Volcanic activity increased, depositing lava flows throughout the northern Sierra. These can still be seen today in many parts of the Tahoe Basin such as Castle Peak, Mount Rose, Tinker's Knob, and Carson Pass. As a result of the change in plate movement, deep north-south cracks known as *faults* developed in the earth's crust, releasing great subsurface forces capable of displacing entire land masses. The release of these forces accelerated the uplift of the Sierra, an elevation rise that continues today. The greater uplift on the eastern side caused the range to tilt westward, creating the sloping western foothills and the steep escarpment that separates the eastern crest from the Great Basin desert below.

The Tahoe Basin formed as a result of the uplifting of two parallel fault blocks, the Sierra and the Carson Range. Subsidence of the crust between these two blocks and subsequent lava flows that dammed outlet streams in the north created

Lake Tahoe, the second deepest lake in North America. Over the last million years glacial outflows of rocks and sediment have supplanted the eroding volcanics to retain the natural barrier that encloses the lake at roughly its present altitude; an artificial outlet dam at Tahoe City maintains this level at approximately 6,300 feet. Intermittent glacial activity also contributed to the formation of the landscape surrounding the lake, scouring the looser volcanic deposits and soil in many areas down to the underlying granite. The impact of this activity can best be seen in areas such as the Desolation Wilderness, southwest of the lake, where subsequent soil build-up has barely begun to cover the huge expanses of open rock left behind by receding glaciers a mere 8,000 years ago.

The climate of California has also changed considerably. Some twenty-five million years ago California was characterized by moderately warm conditions with plentiful rainfall, supporting a broadly diverse mixed coniferous and hardwood (angiosperm) forest, which ranged from the coast to what is now the Great Basin desert. Around this time, for reasons still not completely understood, climatic conditions shifted toward greater temperature extremes and less overall precipitation. Particularly important was the gradual development of the *Mediterranean* climatic pattern, in which summer rainfall is sparse or nonexistent. This climatic change eventually led to the extinction in California of many hardwood species today characteristic of the eastern United States, such as beech and sweet gum, and the ecological isolation of other species, such as the giant sequoia, to specialized habitats where overall moisture was still sufficient for survival.

The effects of California's changing climate were exacerbated toward the east, where the rise of large mountain ranges resulting from uplifting fault blocks formed *rainshadows*, behind which precipitation was greatly reduced. New subalpine and alpine environments were created, characterized by low

The volcanic ridges that loom above the glacially scoured granite in Shirley Canyon are typical of far northern and southern locations in Tahoe.

temperatures and often large amounts of winter snowfall. These conditions were intensified by the coming of the ice age three million years ago, which brought glaciers to the high country of the Sierra Nevada. Since that time mountainous areas in California, including the Tahoe Basin, have been undergoing periodic glaciation, to varying degrees, every 10,000 years or so.

These geologic and climatic forces have combined to form Tahoe's unique and varied environment. Today the Basin is characterized by warm summers with little precipitation, cold nighttime and winter temperatures, frequently heavy winter snowfall, and a variable, sometimes nonexistent, soil base.

Above lake level, Mount Rose in the north and Freel Peak and Job's Sister in the south reach almost 11,000 feet in height. Many other ridges and peaks along the western crest, including Mount Tallac, Pyramid Peak, Dicks Peak, and Castle Peak, rise well above 9,000 feet. The remnants of metamorphic and volcanic peaks and ridges, having thus far resisted the intervals of glacial scouring, lie interspersed in a granitic landscape dotted with glacial lakes and occasional deep sediment deposits. Narrow snowmelt-fed streams rush down steep slopes, accumulating vegetative cover and soil as they expand out into gently sloping, forested watersheds such as Ward Valley, Fallen Leaf Lake, Alpine Meadows, Spooner Lake, and Meeks Bay.

Diversity and Distribution of Tahoe's Flora

Tahoe's floral diversity can be attributed in part to its central location between the plant communities of the western foothills, the eastern Great Basin desert, and the northern and southern Sierra. Many Tahoe plants are restricted to certain locations, such as the northern or southern regions, the Desolation Wilderness, the western lake shore, or the drier peaks of the Carson Range. The variety of the Basin's plant species can further be explained by the area's rapidly changing and varied habitats, which have contributed to the large number of Tahoe plants with limited geographic ranges. In this respect, Tahoe is a microcosm of California, whose fluctuating climate and terrain have made the state a haven for *endemics,* plants found nowhere else in the world. Some, such as the giant sequoia or the coastal cypresses, are ancient relicts, having previously possessed much larger ranges. Others, such as the many endemic clarkias, buckwheats, and sunflower family members, are products of recent, and undoubtedly still ongoing, evolutionary changes.

In appreciating Tahoe's floral diversity, it is worthwhile to consider the mechanics of plant speciation, that is, how plant species become taxonomically distinct from one another. Speciation is generally a product of a random gene mutation, which, in rare circumstances, offers an organism some kind of advantage in producing offspring. If mutated individuals are reproductively isolated from the parent population, varying natural selection pressures can lead over time to new species of plants.

Reproductive isolation may occur through geographic separation of different populations or through the creation of barriers to gene exchange within the same population. In the Sierra the isolation of peripheral populations through climatic or topographical changes was probably a common occurrence in geologic history. This phenomenon can be observed in the many eastern Sierra or Great Basin desert species that occur over limited ranges in the southern and northern parts of Tahoe, a probable result of migrations through corridors that existed thousands of years ago during warmer, drier periods. The advance and retreat of glaciers, in particular, probably fragmented many plant populations. Many species within Tahoe's alpine plant community, for example, are thought to have evolved into distinct taxa as a result of being stranded atop *nunataks*—isolated, windswept summits and ridges that remained free of year-round snow cover during periods of high glaciation.

Reproductive isolation within the same population can occur in a number of ways. Gene mutations may create physical or chemical barriers to reproduction. Species in the same genus may occasionally interbreed, producing viable *hybrid* offspring. Chromosome alterations, particularly the doubling of chromosomes during meiosis to create *polyploid* individuals, may also create gene-flow barriers. When hybrid or chromosome-altered plants intrabreed, or reproduce asexually (as is common in many Tahoe species), the possibility of a new species lineage arises. Chromosome alterations, in particular, have played an enormous role in the radiation of angiosperms, accounting for over fifty percent of all flowering plant species.

VEGETATION ECOLOGY OF THE TAHOE BASIN

Even the most casual observer will notice distinct patterns of vegetation across the mountains and sloping watersheds of the Tahoe Basin. Trees are sparse or even absent on the high ridges. Some mountainsides are full of trees, while others are open and rocky or covered with a thick layer of shrubs. As one moves down a watershed, the vegetation becomes denser and noticeably different in appearance. How and why Tahoe plants occur where they do is the subject of this section.

To understand Tahoe's vegetation ecology, it is helpful to review some of the basic physiological processes plants undergo to survive in the Basin. Plants create the energy needed to grow and reproduce by combining sunlight and carbon dioxide in a process known as *photosynthesis*. Like most living organisms, plants use oxygen to conduct *cellular respiration*. Plants also need nutrients such as phosphorous and nitrogen to build essential organic compounds. Finally, plants require water in order to carry on cellular functions and to maintain pressure in cell walls. A limit on any of these inputs may determine whether a plant can grow in a given habitat. Light, for example, plays an enormous role in determining seedling survival under chaparral or forest canopies, particularly for tree species, such as pines, that require direct sunlight within a few years of germination. Nutrient limitations are also an important factor in plant distribution; this is discussed more fully in the sections below covering soil and biotic relationships.

In the summer-dry climate and rapidly draining soils of the Basin, limits on available sources of water often pose the greatest challenge for young seedlings to become established. Drought-adapted species all have the ability to slow water loss by closing the narrow molecular openings in their leaves known as *stomata*. In closing their stomata, however, plants also impede their intake of carbon dioxide, thereby shutting down photosynthetic activity. Unable to photosynthesize and lacking water, a plant will soon perish in rising summer temperatures that drive its energy demand beyond whatever carbohydrate reserves it may be storing. The relative abilities of plant species to access water sources and/or to carry on photosynthesis in Tahoe's summer environment is a major factor in determining vegetation patterns within the Basin. The following sections discuss the ecological characteristics of Tahoe's environment, how plants have adapted, and the resulting plant communities found in the Basin.

Factors Affecting Vegetation Ecology

Vegetation ecology begins with the recognition that a plant must obtain nutrients and water, find shelter, and reproduce, all while anchored to the same patch of land on which it began life as a seedling. A plant's ability to survive and flourish in its environment thus depends in part on a set of factors that determine the conditions under which the plant germinates. These factors fall into four main categories, which are treated in separate sections below: climate; soils; topography; and biotic relationships. These four factors do not exist in isolation. Instead, each influences the other. Climate, for example, profoundly affects soils through weathering action and is probably also the most important determinant of biotic relationships within a plant community. The local biota, in turn, strongly affects soil character and, often, the microclimate of the immediate environment. The interactions among ecological factors change over time, often in a relatively predictable sequence known as *succession*, which also is discussed below. The concept of succession transforms the snapshot image of the Basin landscape into a moving picture, as vibrant and alive as the plants that play the starring roles, moving inexorably forward together in an ever-renewing cycle of life.

Climate

The main climatic elements influencing vegetation patterns in the Tahoe Basin are temperature, precipitation, and wind. Average annual temperatures in the Sierra tend to decrease up to five degrees for every 1,000-foot gain in elevation. From lake level up to the summit of Mount Rose, this means a potential temperature differential of more than twenty degrees. Because Sierra plants typically are dormant in winter, average daytime temperatures during the summer are more important in affecting a plant's distribution.

Each plant species has its own optimum temperature for photosynthesis. As one moves in either direction from this optimum, a species' ability to survive decreases. Extremely warm temperatures, for example, can limit the distribution of non-adapted species, both by lowering photosynthetic rates and by increasing moisture stress through evapotranspiration. Cold average temperatures can also decrease photosynthetic rates to the point that many plants are unable to produce sufficient energy to conduct cellular respiration. Studies in the Sierra have shown that trees typically will not grow above elevations at which the average daytime temperature in July is lower than fifty degrees Fahrenheit. In contrast, most alpine plants have the ability to photosynthesize at near freezing temperatures.

To moderate temperature conditions, many chaparral shrubs or herbs such as pussypaws (*Calyptridium umbellatum*) respond physically, rotating their stems or leaves to either intercept or reduce sunlight energy. Some botanists hypothesize that mid- to late-season floral color changes from white or yellow to red among plant groups such as the buckwheats (*Eriogonum* spp.) represent adaptations that increase absorption of ultraviolet light, thereby warming the inflorescences and allowing for higher rates of seed maturation.

Freezing nighttime temperatures in summer also may affect plant growth by damaging cellular tissue. Although most plants in the Basin are adapted to summer frosts, the frequent and often rapid onset of freezing temperatures characteristic of subalpine and alpine environments undoubtedly limits the upper distribution of many lower-elevation species.

Precipitation is another key factor affecting plant distribution. In Tahoe, most precipitation falls as snow from winter storms out of the west. As eastbound storms rise to meet the western crest of the Sierra, they deposit most of their moisture. The Sierra rainshadow causes yearly precipitation to decrease generally across the Basin as one drops in elevation and moves eastward. Along the western ridges of the Tahoe Basin, from Donner Summit in the north to Echo Summit in the south, precipitation may reach eighty inches in good moisture years, significantly higher than at the elevation of the lake and more than twice the amount found along the eastern shore. High-elevation outposts such as Mount Rose or Freel Peak receive disproportionate amounts of moisture relative to their surrounding areas. As one descends the east side of the Carson Range, outside the Basin, a second rainshadow reduces yearly precipitation below twenty-five inches, the amount required to support tree growth.

In Tahoe, the timing and nature of precipitation, in conjunction with soil type and topography, strongly influence the length of a plant's growing season. In the foothills mild temperatures allow for plant growth to accompany fall and winter rains, culminating in springtime flowering followed by dormancy once soil moisture is depleted. In the high Sierra, colder winter temperatures, frozen ground, and deep snow delay the start of the growing season into summer. As snowfields melt and drain through the soil, plants take up as much moisture as possible. Occasional late spring and summer thunderstorms prolong the growing season by replenishing the dry upper soil layers.

Tahoe plants cope with summer drought conditions in different ways. Some, such as Jeffrey pine or green-leaf manzanita (*Arctostaphylos patula*), prefer dry, well drained, granitic soil, in which their roots may reach down thirty feet to access groundwater. In addition, pines and manzanitas have small, leathery leaves, which are covered by a waxy outer layer known as a *cuticle*. Thick cuticles reduce water loss during periods of high temperature or moisture stress. Cuticles also protect evergreen plants against moisture loss through evapotranspiration during winter, when frozen groundwater is inaccessible to plant roots. The evergreen condition, shared by Tahoe's conifers and most of its chaparral shrubs, allows plants to exploit early or late-season opportunities for photosynthesis when sunlight and groundwater are both readily available. Evergreen plants also save energy each spring by using the same leaves from the previous season.

An important additional factor influencing the length of the growing season is snowpack depth and exposure. In north-facing, closed-canopy forests, snow melts slowly, often remaining on the ground well into July, thereby providing a summer-long supply of moisture for growth. Lingering snowpacks, however, also can inhibit plant distribution. In meadow environments, soggy conditions prevent plant roots (particularly those of trees) from obtaining sufficient oxygen to commence growth. On high-elevation northern and eastern exposures, snowbanks may last all summer, preventing the growth of snow-covered seedlings or perennials, as well as plants downslope from the cold, soggy runoff.

The lack of deep snowpack at the level of the lake, particularly on the eastern side, also may influence plant distribution by intensifying summer drought, leading to a less diverse assemblage of plant species. Moisture stress is particularly acute on high, wind-swept ridges and peaks, where the thin snow layer quickly melts with the onset of warm temperatures, signaling the start of a race against time for the small alpine perennials to grow, flower, and set seed before soil moisture is depleted. Wind exacerbates these conditions, greatly increasing evapotranspiration and reducing the daily photosynthetic level of each plant.

Snow cover also plays an important role in the distribution of plants at higher elevations. In these environments snow envelops plants in a protective shield against freezing winter temperatures and hard, wind-blown ice crystals. While timberline elevation is influenced primarily by temperature conditions, snow depth will often determine the height that *krummholzed* trees or shrubs may attain. A good example of

Krummholz forms of whitebark pine grow to the height of the average winter snowpack on Freel Peak.

this phenomenon may be seen on the southern ridges of Freel Peak, where whitebark pines grow as matted shrubs in a gardenesque manner, neatly trimmed at the level of the usual winter snowpack. The harshest environments in the Basin are found on the open ridges and peaks, where gale-force winter winds keep the rocky terrain relatively free of protective snow cover, allowing for survival of only the smallest matted perennial herbs.

Soils

Soil is the medium through which a plant obtains moisture and nutrients, oxygen for its roots, and support for its vascular structure. Soils vary in chemical composition, average particle size, drainage, and depth, and each of these factors can strongly affect plant distribution.

To survive, plants rely on a number of essential nutrients, including nitrogen, phosphorous, calcium, potassium, and magnesium. A soil's chemical composition determines, in part, which nutrients will be available. The soils of the Tahoe Basin generally are nutrient-poor, especially near the surface where drainage greatly exceeds the rate of chemical erosion from individual soil particles. Plants cope with nutrient deficiency through mutualistic relationships with mycorrhizal soil fungi or through associations with plants capable of nitrogen fixation. The marsh-loving sundew plant (*Drosera rotundifolia*) obtains nitrogen by capturing small insects in its sticky glands and slowly digesting them with secreted enzymes.

Tahoe soils vary considerably in average particle size. For example, the fine clay soils typical of meadows or other sediment deposits may have an average particle size less than one-hundredth that of gravelly soils characteristic of many subalpine slopes. Soils with large average particle size have better drainage and oxygen capacity, but are less able to retain moisture and nutrients. Smaller particle size inhibits drainage and reduces oxygen capacity, but allows for greater retention of moisture and nutrients. Many plants are unable to survive the moisture stress associated with the rapidly draining sandy or rocky soils found throughout the Basin. Conversely, where poor drainage results in waterlogged soils, as is common in wet meadows or alpine depressions, most plants have a difficult time growing because of the lack of oxygen.

Tahoe soils also differ greatly in depth. On the glacially scoured terrain of the Desolation Wilderness, as little as six inches of soil may have developed above impermeable bedrock. When soils are this thin, annual herbs are often the only species that can survive, which they do by growing quickly and producing seed before soil moisture is depleted. The best soils in the Basin, where erosion sediments have accumulated in deep, well drained layers, typically are occupied by moisture-loving forest species such as white fir or red fir.

Soils are derived from two essential sources: parent rock and organic material. Organic matter is composed of plant debris discarded on the surface or old root material in the soil. As one digs deeper, organic material disappears. Plant litter replenishes the soil with nutrients that would otherwise have leached out of the upper soil layers. Organic material also helps to break up compacted soils. However, plant debris, particularly conifer needles, may alter the pH of soil, often rendering it highly acidic and thus inhospitable to many plants.

Parent rock in the Basin is usually composed of grandiorite, a form of granite, or andesite, a volcanic rock. More susceptible to chemical breakdown, volcanic soils tend to contain more nutrients than granitic soils, which are often nutrient-poor. Granite and volcanic soils also vary in particle size, drainage, and depth of root access. Having formed slowly under the earth's surface, granitic rocks are dense and highly resistant to erosion. Volcanic rocks, having been spewed forth into the atmosphere and cooled quickly, are less dense and erode more readily. Thus, volcanic soils have higher clay content and drain more slowly, thereby providing preferred habitat for plants with superior abilities to access water sources close to the surface. In contrast, granitic soils are typically sandy, with relatively large average particle size and quick drainage. Where the underlying granite bedrock is older, fractures may occur, allowing for deep root penetration. These conditions increase the potential for moisture stress near the soil surface, but provide optimum habitat for plants that can access groundwater sources through well developed taproot systems. In the mid- to late-summer heat, deeply fissured granitic soils actually may offer better sources of moisture than the shallow soils characteristic of many high-elevation volcanic habitats in the Basin.

Topography

One need only glance at two opposing mountainsides, one covered by shrubs and the other by dense forest, to appreciate the often overlooked influence of topography on plant distribution. The direction a slope faces in relation to the sun, or its *exposure*, will determine the amount of light energy it receives. In the northern hemisphere, southern exposures receive significantly more sunlight than northern ones. Western exposures, while sharing equal sunlight with eastern exposures, receive light in the later, warmer part of the day. As a consequence, southern and western slopes in the Basin receive more heat and tend to be drier than northern and eastern slopes. In summer prevailing southwest winds may intensify drought conditions by increasing evapotranspiration and moisture loss. Winter winds reduce snow cover on higher, western-facing slopes by blowing snow over the ridge, where it piles up on northern and eastern exposures, forming dense snowpacks that last into the summer.

Differences in temperature, moisture, and light due to variations in exposure can result in distinct vegetative patterns. Many tree species, such as red fir or mountain hemlock (*Tsuga mertensiana*), are confined mostly to cooler northern or eastern exposures. These moister slopes often support luxuriant forest growth, interspersed with shade-adapted shrubs and herbs as well as various non-photosynthesizing species of the heath and orchid families. Over the ridge, southern or western exposures typically are covered with montane chaparral, with an intermingling of dry-adapted trees and herbs.

Topographical features of the landscape also play a key role in determining moisture availability. Some of the driest habitats in the Basin are steep southern exposures with rapid runoff and little percolation of snowmelt. Many exposed volcanic plateaus and slopes in the Basin offer a significantly drier mid- to late-summer habitat than more shaded granitic outcroppings, despite the generally superior moisture retention of volcanic soils. In contrast to these moisture-stressed environments, depressions or riparian channels allow water runoff to form permanently moist habitats that host an array of plant species with no special adaptations to summer drought. Such habitats, common to areas such as Sagehen Meadows, Pole Creek, Shirley Canyon, Paige Meadows, Osgood Swamp, and Carson Pass, may range from alpine creeks to mossy seeps, wet or moist meadows, and forested riparian waterways. Wherever there is a season-long water supply, one will be sure to find the most extravagant wildflower blooms.

Biotic Relationships

A plant's ability to survive and reproduce in a given habitat may be greatly affected by its interactions with other living organisms. These biotic relationships may be organized into four general categories: predator-prey relationships, mutualistic relationships, parasitic relationships, and competitive relationships. Each category is discussed below.

Predator-prey relationships. Many types of organisms, from bacteria to deer, rely on plant parts as a food source. Plants have developed a varied assortment of defenses against these predators. Mountain whitethorn (*Ceanothus cordulatus*), various gooseberries (*Ribes*), and mountain rose (*Rosa woodsii* var. *ultramontana*) all have sharp spines that discourage hungry foragers. Stinging nettle (*Urtica dioica*) bears small needle-like hairs that release formic acid

Herbivorous rodents such as this golden-mantled ground squirrel may play a significant role in influencing vegetative composition.

upon contact with any unlucky intruder. Other plants, such as those in the mint family or in the sunflower genera *Artemisia* and *Madia*, produce chemical compounds in their leaves and other plant parts that discourage herbivory. Conifers produce sap, a sticky, sugary liquid that overwhelms their many insect predators, particularly the various species of bark beetle, genus *Dendrocotonus*. Trees that are less healthy, often due to long-term moisture stress, do not produce sufficient amounts of sap and thus become susceptible to beetle attacks.

The degree to which herbivory affects the distribution of plants in a given area is difficult to gauge. It is possible to imagine, however, that predation may often preclude any one plant species from exclusively occupying a given habitat. The greater abundance of aspens (*Populus tremuloides*) on the eastern side of the Basin has been attributed in part to the higher concentration of herbivorous rodents on the western side, where rainfall is more plentiful. A late-summer stroll through a wide patch of large-leaf lupine (*Lupinus polyphyllus*) may reveal many pea-like fruits, each infested with various insect larvae, none of which will produce viable seed. The spectacular radiation of the sunflower family may have been due in part to the unique manner in which the sunflower inflorescence encases each seed-producing ovule in a protective ovary, thus vastly increasing the odds that many of its seeds will escape insect predation.

Mutualistic relationships. Mutualistic relationships in ecology are defined as interactions that benefit both of the organisms taking part in the interaction. A simple example is the relationship between many plants and fruit- or seed-eating animals. Seeds contained within the berry-like cones of Sierra juniper (*Juniperus occidentalis*), for example, must pass through an animal's digestive track in order to germinate. The parasitic mistletoes (*Arceuthobium* and *Phoradendron* spp.) rely on birds to eat their berries and then deposit still viable seeds, with accompanying fertilizer, on the branches and trunks of host trees.

Perhaps the best known mutualistic relationship is the coevolution of flowering plants and their pollinators. A major trend in this coevolution has been the formation of increasingly specific relationships between certain plant genera and more efficient pollinators. Bees are highly desirable pollinators because of their ability to recognize, and thus return to, the flowers of the same plant species. The pollination strategies of many Tahoe plants such as *Penstemon*, *Lupinus*, or the various genera that are "buzz pollinated" reserve pollen and nectar only for bees. These flowers often have faint nectar lines or spots that are highly visible to bees, whose vision is oriented toward the ultraviolet end of the color spectrum.

Color, structure, and smell may each play a role in determining how a flower is pollinated. Long, tubular corollas, characteristic, for example, of many species in the phlox family

(Polemoniaceae), often reserve nectar for the long, coiled proboscises of butterflies and moths. Predictably, moth-pollinated flowers such as the Washington lily (*Lilium washingtonianum*) are light in color and often strongly scented to aid in nighttime location. Close inspection reveals that red flowers in the Basin, such as the unrelated scarlet gilia (*Ipomopsis aggregata*) and California fuchsia (*Epilobium canum*), tend to be long, tubular, and odorless. These flowers reserve nectar for hummingbirds, which have a poor sense of smell but, unlike insects, see the color red distinctly. Other flowers, such as spreading phlox (*Phlox diffusa*), change color after pollination, thereby directing prospective visitors to still fertile flowers. In some species only certain flower parts change color, such as the flower tube throat or the spot on the banner petal of various lupines.

A lesser known but ultimately more essential mutualistic relationship is the interaction between plants and mycorrhizal fungi. In the mycorrhizal relationship, plant roots are penetrated by networks of underground fungal filaments that cause the roots to branch and spread, thus increasing their access to water and nutrients. This relationship provides the fungi with essential carbohydrates and the plant with scarce resources needed for continued growth. It is believed that the mycorrhizal condition was an essential factor in allowing plants to make the original transition from aquatic to terrestrial environments millions of years ago. In the nutrient-deficient soils of the Sierra, mycorrhizal interactions help to determine where plants can survive. While most plants have at least some beneficial contact with soil fungi, certain groups, such as the conifers or the varied genera in the heath family, benefit from highly specific mycorrhizal relationships that give them competitive advantages over other species that exchange resources less efficiently.

Another important mutualistic relationship occurs in several families of plants that form root nodules inhabited by specialized bacteria. In exchange for a steady supply of carbohydrates, the bacteria convert atmospheric nitrogen into compounds that the plant can use for growth and to maintain cellular activity. Nitrogen fixation is a crucial component of many plant communities in the Sierra, where nitrogen is almost always in short supply. Symbiotic relationships with nitrogen-fixing bacteria are common among the widespread members of the pea family, which includes locoweeds, lupines, wild pea, clovers, and lotus. These relationships are also known to occur in alders and ceanothus.

Parasitic relationships. Parasitic relationships are defined as interactions in which one organism (the parasite) benefits while the other (host) organism is harmed. Some parasitic plants, such as members of *Castilleja, Orthocarpus,* or *Pedicularis* in the figwort family, have the ability to photosynthesize but will supplement this production by taking water, nutrients, and sugars from nearby host-plant root systems. The photosynthetic mistletoe extracts water and nutrients from the host plant, while also using it for structural support. Others, including California dodder (*Cuscuta californica*) and members of the broomrape family (Orobanchaceae), no longer possess chlorophyll and so have lost their ability to photosynthesize, relying instead entirely on their hosts to obtain the products of sunlight energy.

A complex and still not completely understood quasi-parasitic relationship is the interaction between soil fungi and *saprophytes,* which have been traditionally defined as non-photosynthesizing plants that obtain nutrients and carbohydrates from decaying forest litter. These include Tahoe species such as spotted coral root (*Corallorhiza maculata*) and various monotrops in the heath family, the best known of which is the bright red snow plant (*Sarcodes sanguinea*). Recent research has demonstrated that these plants rely on soil fungi to obtain nutrients and carbohydrates from decaying organic material. In addition, some species are thought to siphon off resources from mycorrhizal fungi that associate with photosynthesizing forest plants, particularly the conifers.

Many types of fungi regularly parasitize plants. The best known is the non-native fungus that causes white pine blister rust. Blister rust affects conifers in the white pine group, which includes all the five-needled pines in the Basin. The fungal spores begin their lives in the leaves of shrubs in the gooseberry family (the infection can be observed as bright scarlet spots on the leaves), where they grow over the summer, then in fall produce a second set of spores that travel by wind to nearby white pines. The second-generation spores invade the bark of the pines and spread around the trunk, eventually killing the trees, but not before producing a new round of spores, which begin the cycle anew by invading neighboring shrubs. Other negative effects of fungal parasitism can be observed in early summer among various species of rock cress (*Arabis*), an often erect herb in the mustard family. Unlike non-infected plants, those parasitized by the rock-cress fungus are relatively short, have orange-colored leaves, and usually bear no flowers.

Insects often choose plants as hosts for the development of their larvae. As is true for most *Quercus* species, for example, huckleberry oak is regularly visited by small wasps that lay their eggs in the plant's leaves or stems. Once hatched, the larvae excrete a substance that causes the oak to produce specialized tissue, which grows around the larvae in a mottled, greenish brown sphere called a gall. Galls provide protection and a source of food for the developing insect. Similar interactions occur between wasps and many other Tahoe plants, including mountain sagebrush and various willow species.

Competitive relationships. A fourth biotic interaction that affects plant ecology is competition. Competition may be *intraspecific*, meaning that it occurs within the same species, or *interspecific*, between different species. A good example of intraspecific competition is the process of thinning, in which a group of same-species plants is gradually reduced in population as more advantageously situated individuals begin to monopolize the resource base to the exclusion of their kin.

There are two types of interspecific competition. In *interference competition*, one species directly prevents another from obtaining available resources. This is a common event in the animal kingdom, in which many species actively exclude others from their territories. Plants are known to engage in this activity through *allelopathy*, a chemical process in which a mature plant exudes toxins into the soil that inhibit the germination of potential competitors. Some researchers have postulated that the sometimes uniform distribution of woolly mules ears (*Wyethia mollis*) or sagebrush (*Artemisia tridentata*) in certain habitats is caused by allelopathy. Conifers also engage in interference competition by littering the forest floor with a thick layer of needles. Needle litter decomposes slowly, and the resulting duff may prevent many germinating seedlings from reaching the soil, or if buried, from reaching the surface. Decomposing needles are also highly acidic, and the resulting change in the soil pH prevents many plants from growing successfully. In an analogous fashion, grasses, sedges, and rushes have fibrous root systems that form tangled mats of underground growth in wet and dry meadows throughout the Basin, inhibiting the germination of other plants.

A second type of interspecific competition occurs when two species compete for essential resources such as water, sunlight, or nutrients. When one species is able to deplete a resource to a level below that necessary for another species to survive, the first species is said to have *competitively excluded* the second from the habitat. One way to understand competitive exclusion is to consider the frequent discrepancy between a plant's potential habitat range (the environment in which a plant could survive) and the area in which a plant actually occurs. Many plant species, for example, undoubtedly would thrive in the deep soils and protected basins characteristic of the red fir forest. Instead, this community is one of the least diverse in the Basin because sunlight, moisture, and nutrients are monopolized by the firs.

Since plants generally compete for the same resources, one might assume that a model based on competitive exclusion would result in a vegetative distribution dominated by a few superior competitors. Instead, even casual observation of the Tahoe landscape reveals a diverse assortment of coexisting plant species.

Species coexist for several reasons. First, plants compete not for just one, but for a combination of resources. Red firs, for example, are shade-tolerant as young trees, and thus may persevere in the forest understory until finally forming their own dense canopy. Red firs require relatively moist soil conditions, however, and thus do not photosynthesize and grow as efficiently as other drought-adapted conifers in drier soils. In intermediate moisture conditions, individual trees of red fir may become established, but a dense, sun-blocking canopy will not develop, and a more diverse growth of pines and various understory shrubs and herbs will occur.

A second reason for species coexistence is that plants are limited by a host of environmental conditions such as temperature or soil character that, in contrast to resources, cannot be reduced or significantly altered through consumption. While environmental conditions apply relatively equally to all plants, species vary tremendously in their ability to access resources under the same set of conditions. Tahoe's pine tree species are illustrative of the many plants that flourish because of their capacity to tolerate marginal environmental conditions that other plants cannot. Lodgepole pines, for example, have the ability to control water uptake in such a way that they can survive in alternating soggy and dry areas that would be intolerable for other conifers. Jeffrey pine is abundant throughout

Tahoe's pines are abundant on the rocky, shallow soils common in the Desolation Wilderness, as viewed from Ralston Peak.

much of Tahoe's lower-elevation habitat through its tolerance for warm, dry conditions and rocky soil. Studies have shown that Jeffrey pine, and its lower-elevation cousin, ponderosa pine (*Pinus ponderosa*), can continue to photosynthesize in higher temperatures and under greater moisture stress than either of the native firs. Whitebark pine (*Pinus albicaulis*) can tolerate cold, windy alpine environments that are largely uninhabitable for other Tahoe tree species.

The different pressures and opportunities available through interactions with other organisms also influence species coexistence. Predation, particularly by insects, undoubtedly prevents many otherwise superior competitors from monopolizing certain landscapes. Conversely, parasites such as coralroot may coexist with red fir in its shady habitat by dispensing with photosynthesis and obtaining essential resources from fir-fungus mycorrhizal interactions.

Finally, species coexistence occurs as a result of environmental change over space and time. Large-scale geographical variation in resource availability and environmental conditions is evident to any Basin hiker who crosses from a shaded stream course onto a sun-drenched, south-facing mountainside or from a protected, north-facing slope onto a windy, rocky summit. Less evident to the casual observer are the minute changes in terrain that allow for variations in resource availability within a small area. Outside the denser forest communities, microhabitats abound: small depressions where snow runoff collects, isolated pockets between boulders that allow for uncontested sunlight and protection from the wind and cold. Slight differentials in resource availability and environmental conditions can result in high plant diversity within a small area.

Succession

Resource availability can also change in the same location over time. This phenomenon, in which different plant communities occupy a site in a relatively predictable progression of stages, is called *succession*. Early successional stages may lead to what is referred to as a *climax plant community*, which, absent some new disturbance, will not change in general species composition. The red fir forest is an example of a climax plant community.

Primary succession occurs when pioneer species occupy a newly formed habitat that is essentially devoid of life. Much of the vegetative history of the Tahoe Basin since the last glaciation approximately 10,000 years ago is one of primary succession, from barren, glaciated landscapes to luxuriant forests. Today, primary succession is still occurring in the high country of the Basin, where rocky outcroppings are slowly colonized by early pioneer plants in cracks and fissures containing accumulated wind-blown dust and organic debris, the rudimentary bases for soil development.

With the conspicuous exception of Lake Tahoe, most lakes in the Basin, such as those that dot the landscape of the Desolation Wilderness, are products of past glaciation. These lakes represent transition stages between primary succession and climax forest. Higher-elevation lakes are rocky, with little soil and few plants. Lower down, lakes accumulate sediment deposits, with corresponding increases in aquatic and shoreline plants. Increased sediments and organic debris produce ever shallower waters, giving the lake a more swamp-like appearance. As this process continues, wet meadows form on the lake's periphery, gradually filling in toward the middle. On the drier edges, woody shrubs such as willows or various members of the heath family take hold, followed by lodgepole pines and then red or white firs. Each group of new vegetation, by stabilizing the soil and adding organic litter, paves the way for the next successional stage. Paige Meadows, in the northwest portion of the Basin, is a good example of a former lake making the transition from wet meadow to pioneer forest community.

Secondary succession occurs as a result of a lesser disturbance, such as flood, fire, or treefall, to an existing vegetative community. The majority of lower-elevation forests in the Basin are still in the process of secondary succession as a result of widespread logging and subsequent fires at the turn of the century. Fire is a common disturbance in the Sierra, and many plants in the Basin are adapted to it. The seeds of manzanita and ceanothus may not open until they have undergone the scarifying heat of a ground fire. Lodgepole pine cones open profusely following fire, quickly colonizing newly burned areas. Fires expose forest or shrub understories to sunlight, while simultaneously laying down a nutrient-rich layer of ash, providing ideal growing conditions for young pine seedlings and other shade-intolerant shrubs and herbs. In the years immediately following a fire, herbs are plentiful, often germinating from long dormant seeds to produce showy flower exhibitions on the open terrain.

In the forest, cool ground fires historically cleared out understory vegetation while preserving the larger trees to produce seed for the newly created habitat. Fire suppression, an official public policy during the twentieth century, has prevented the creation of sunlit forest openings, thus favoring shade-tolerant species such as firs or incense-cedar at the expense of pines. The crowded understories of these forests create the possibility for hotter crown fires that can eliminate mature trees, leaving behind the type of devastated landscape that resulted from the 1988 fire in Yellowstone National Park. Forestry officials in the Tahoe Basin are now confronted with the problem of how to remove dense undergrowth to restore diversity to the forest environment, while avoiding an out-of-control conflagration.

The concept of plant succession leading to a stable climax community has been criticized as portraying vegetation ecology as a uni-directional process with an ultimately static, and implicitly superior, end goal. Most plant ecologists now acknowledge that vegetation patterns are dynamic, that community disturbance is both regular and, viewed on a broader time scale, frequent, and that plant species have adapted to this reality in different ways.

Some plants characteristic of climax communities are adapted only to particular habitats, such as freezing summit peaks or shady forest floors. By creating a unique niche for themselves, specialists forego the chance to live in other habitats. Other plants, such as woolly sunflower, pussypaws, and lodgepole pine, may possess general adaptations applicable to a wide variety of habitats. In their preferred habitat, specialists usually outcompete generalists, but generalists may prosper by occupying marginal or newly created habitats. As befits their opportunistic nature, generalists typically have a potential range that greatly exceeds their distribution.

Plants ultimately bring a host of capabilities to the competitive arena, each of which has played some small or large part in ensuring the continued survival of the species. The variation in plant adaptations illustrates how complex it can be to evaluate interspecies competition over time. Early successional *pioneer species*, such as soil-stabilizing lodgepole pine, or nitrogen-fixing ceanothus or lupine, may create favorable conditions for later-stage species such as firs or manzanita to occupy a given environment. In a sense, these later successional species outcompete earlier pioneer species. From a different perspective, however, the various species can be seen as taking turns occupying a continually changing terrain. Many scientists believe that evolutionary diversification has resulted more from the creation of new habitats than from direct competition. Most of Tahoe's flora undoubtedly diverged along distinct evolutionary paths long before the most recent rise of the Sierra and the onset of glaciation intervals several million years ago. Thus, whatever competition may have influenced evolutionary paths, most of it occurred before the existence of the Tahoe Basin habitat as we know it today.

Vegetative Communities of the Tahoe Basin

Vegetative communities are identified by common vegetation associations and patterns, often with an associated elevation range. Communities in the Basin occur over three of the oft-described *life zones* of the Sierra Nevada: the upper montane, subalpine zone, and alpine zone. Because of the varied landscape that characterizes the Sierra Nevada, botanists do not agree on the exact number of plant communities in the region. Given the honest differences of opinion regarding the importance of small variations in habitat, such a consensus probably would add little to our understanding of the regional ecology. As in taxonomy, the concept of plant communities offers insight into natural systems by presenting an overlying structure that may or may not always correspond to the situation in the field. In considering plant communities, it is useful to keep in mind the distinct potential for randomness in vegetative distribution due to factors such as seed placement or short-term climatic conditions. Plants, like all other organisms, will attempt to survive with the hand they are dealt. Their resilience is a challenge to our capacity to think flexibly and to learn what we can from observable patterns while appreciating the spontaneity of exceptions to the rule.

Upper Montane Life Zone

The upper montane life zone ranges from lake level at 6,230 feet to around 8,000 feet elevation and includes the largest area and greatest variety of habitat in the Basin. At lower elevations within this zone, up to approximately 7,000 feet, the **white fir forest** is found on moister sites. Although white fir is the most abundant tree species, the white fir forest is essentially a mixed community, containing ponderosa pine (at the lowest elevations), Jeffrey pine, lodgepole pine, incense-cedar, sugar pine, and, at slightly higher elevations, red fir. Depending upon variations in moisture availability and topography, this community can form a dense canopy, but it is often relatively open, with a three-layer understory consisting of deciduous trees or large shrubs such as Scouler's willow (*Salix scouleriana*), aspen or mountain maple (*Acer glabrum*), smaller shrubs such as double-flowered honeysuckle (*Lonicera conjugialis*), Sierra

Red firs hug the moister, protected northeastern slopes near Alpine Meadows.

currant (*Ribes nevadense*), Sierra gooseberry (*Ribes roezlii*), or thimbleberry (*Rubus parviflorus*), and many herbs, including sweet cicely (*Osmorhiza chilensis*), crest lupine (*Lupinus arbustus*), Jessica's stickseed (*Hackelia micrantha*), Solomon's seals (*Smilacina* spp.), nodding microseris (*Microseris nutans*), Lemmon's catchfly (*Silene lemmonii*), Nuttall's larkspur (*Delphinium nuttallianum*), and single-stemmed groundsel (*Senecio integerrimus*), among others. The white fir forest is best seen in the Basin on the western shore of Lake Tahoe.

Above the white fir forest, on similarly deep, moist, and well drained soil, one finds the **red fir forest**. This community ranges from 7,000 to 8,500 feet elevation, typically on northern or eastern slopes. These environments commonly receive the largest amounts of snow in the winter. The cooler exposures allow snowpack to remain on the ground longer into the summer, prolonging the growing season. In this protected environment, red firs exploit available moisture and light, creating a dark, almost monocultural environment with only a scattering of single-layer understory vegetation consisting almost exclusively of perennial herbs. Where sunlight peeks through, associates of the red fir forest may include the yellow-flowered Brewer's golden aster (*Aster breweri*), white-flowered hawkweed (*Hieracium albiflorum*), broad-seeded rock cress (*Arabis platysperma*), pinewoods lousewort (*Pedicularis semibarbata*), and ballhead phacelia (*Phacelia hydrophylloides*). Many of these shade-tolerant species are able to photosynthesize at low light levels and prefer cooler conditions to the hot, sunny environment outside the forest habitat. As the forest understory becomes shadier, species in the wintergreen group of the heath family, such as white-veined wintergreen (*Pyrola picta*) and the prince's pines (*Chimaphila* spp.), survive by associating with mycorrhizal fungi. Tapping into the underground flow of nutrients and carbohydrates, non-photosynthetic species such as spotted coral root and the heath family monotrops, notably snow plant and pinedrops (*Pterospora andromedea*), rise out of the needled forest floor. Most of these understory species are also prevalent in shadier habitats within the white fir forest.

Jeffrey pine forest covers drier areas across the Basin from lake level to about 8,000 feet elevation. In this community Jeffrey pines form open stands with occasional sugar pines, lodgepole pines, incense-cedar, and white fir. At higher elevations sugar pine is replaced in the Jeffrey pine forest by its more cold-tolerant cousin, western white pine, and white fir is replaced by red fir. Jeffrey pines are also common throughout the Basin on thinly forested southern exposures, where they create patchy associations with montane chaparral communities and, in rocky areas, Sierra juniper. In these moisture-stressed environments one also finds manzanitas, bitter cherry (*Prunus emarginata*), serviceberry (*Amelanchier utahensis*), mahala mat (*Ceanothus prostratus*), sulfur buckwheat (*Eriogonum umbellatum*), gay penstemon (*Penstemon roezlii*), white-veined mallow (*Sidalcea glaucescens*), and Applegate's paintbrush (*Castilleja applegatei*). On the eastern side of the Basin, where precipitation is lower but soils are still fairly deep, Jeffrey pines are abundant across large open areas with understories composed of nitrogen-fixing lupines and ceanothus, along with mountain sagebrush (*Artemisia tridentata* ssp. *vaseyana*), bitterbrush (*Purshia tridentata*), slender penstemon (*Penstemon gracilentus*), and Brown's peony.

Montane chaparral is recognized as a distinct community beginning in the upper montane and extending into the subalpine zone on rocky, typically granitic southern exposures or as a successional community following fire in red fir forest. Montane chaparral plants possess the typical characteristics of drought-adapted species: small, leathery, often evergreen leaves and deep taproot systems that exploit fissures in the weathering bedrock to access groundwater after surface moisture has disappeared. In Tahoe the montane chaparral community usually consists of huckleberry oak (*Quercus vaccinifolia*), Sierra chinquapin (*Chrysolepis sempervirens*), and manzanita, with cream bush (*Holodiscus discolor*), mountain whitethorn, and tobacco brush (*Ceanothus velutinus)* as other frequent associates. Occasional openings in the chaparral may reveal treasures such as the showy Washington lily or the diminutive dwarf monkeyflower (*Mimulus mephiticus*), whose bright flowers may be either yellow or magenta.

The absence of montane chaparral on open, dry slopes in the montane zone of the Basin is usually attributable to the presence of shallow soils overlying impenetrable bedrock. Where such conditions prevail, one finds an open plant community in which chaparral shrubs are opportunistically interspersed with Sierra juniper, Jeffrey pine, western white pine, and such common perennials as mountain pride (*Penstemon newberryi*), King's sandwort (*Arenaria kingii*), spreading phlox, Sierra stonecrop (*Sedum obtusatum*), hot-rock penstemon

(*Penstemon deustus*), bear buckwheat (*Eriogonum ursinum*), Wright's buckwheat (*Eriogonum wrightii*), mountain jewel-flower (*Streptanthus tortuosus*), and California fuchsia (*Epilobium canum* ssp. *latifolium*).

Because volcanic bedrock tends not to fissure, volcanic soils typically lack the deep groundwater sources offered by granitic outcroppings and preferred by deep-rooted chaparral species. Thus, montane chaparral may also be conspicuously absent on open volcanic slopes, which are instead monopolized by woolly mules ears and mountain sagebrush with balsam root (*Balsamorhiza sagittata*), mountain pennyroyal (*Monardella odoratissima*), mountain snowberry (*Symphoricarpos rotundifolius*), bitterbrush, California valerian (*Valeriana californica*), stickseeds (*Hackelia* spp.), Davis' knotweed, (*Polygonum davisiae*), and Brewer's navarretia (*Navarretia breweri*) as important associates. Two other parasitic plants sometimes found in this volcanic environment are Copeland's owl's clover (*Orthocarpus cuspidatus*) and, at higher elevations, the two dry-adapted broomrapes, *Orobanche corymbosa* and *O. fasciculata*.

Numerous habitats in the upper montane zone may be distinguished from the communities discussed above by the presence of surface moisture for most, if not all, of the growing season. Below 7,000 feet elevation, water often flows in forested riparian channels such as the upper and lower Truckee rivers, Bear Creek, Slaughterhouse Creek, Ward Creek, and McKinney Creek. In contrast to the summer-drought communities found throughout the Basin, riparian plants have no special adaptations to moisture stress, relying instead on a summer-long source of accessible groundwater. Moisture-loving angiosperm trees and shrubs such as black cottonwood (*Populus balsamifera*), willows (*Salix* spp.), aspen, mountain alder (*Alnus incana* ssp. *tenuifolia*), mountain ash, and creek dogwood (*Cornus sericea* ssp. *sericea)* flourish in these environments. Below the

canopy, in alternating patches of sun and shade, live herbaceous perennials such as hedge nettle (*Stachys ajugoides*),
fireweed (*Epilobium angustifolium*), alpine lily (*Lilium parvum*),
common monkeyflower (*Mimulus guttatus*), and American
speedwell (*Veronica americana*). In the stream itself one may
find the water buttercup (*Ranunculus aquatilus*), whose small
white flowers sometimes open beneath the water surface.

In the upper elevations of the montane zone, narrower,
faster stream channels with rocky, well oxygenated substrate
provide vivid wildflower displays. In these environments,
mountain alders and willows, and occasional creek dogwoods
and coffeeberries (*Rhamnus rubra*), form densely shaded tun-
nels that open up regularly into cascading falls and granite
pools lined by perennials such as brook saxifrage (*Saxifraga
odontoloma*), soft arnica (*Arnica mollis*), Brewer's bishops cap
(*Mitella breweri*), alpine speedwell (*Veronica wormskjoldii*),
arrowleaf groundsel (*Senecio triangularis*), Coulter's daisy (*Erigeron coulteri*), enchanter's nightshade (*Circaea alpina*), heart-
leaf bittercress (*Cardamine cordifolia* var. *lyallii*), monkshood
(*Aconitum columbianum*), stream violet (*Viola glabella*), towering larkspur (*Delphinium glaucum*), giant red paintbrush
(*Castilleja miniata*), large-leaf lupine, bog orchid (*Platanthera
sparsiflora*), and Lewis' monkeyflower (*Mimulus lewisii*).

Drainage basins in the upper montane zone exhibit a variety of successional stages between glacially scoured granite
basin and mature forest, which include lakes, swamps, and wet
to dry meadows. Along the shores of Lake Tahoe, one may find
the endangered Tahoe watercress (*Rorippa subumbellata*) surviving amidst other species that have persevered despite increasing human use of the lake shore. As one climbs in elevation to destinations such as Shirley Lake, Five Lakes, or Stony
Ridge Lake, a distinct shoreline community becomes noticeable, composed of heath family members such as labrador tea
(*Ledum glandulosum*) and red mountain heather (*Phyllodoce*

breweri), along with alders, willows, and an overhead canopy of lodgepole pine and encroaching red fir.

In moving from lakes to swamps and meadows, community classification begins to blur in myriad potential categories distinguished by slight variations in hydrologic, soil, and species characteristics. The distinction between a swamp and a meadow is not always clear, and many wetland areas of the Basin, such as Paige Meadows, Osgood Swamp, and Grass Lake, possess elements of both habitat types. In this environment one may find, in the open water, yellow pond-lilies (*Nuphar lutea* ssp. *polysepala*) and common bladderwort (*Utricularia vulgaris*); in the shallows, water smartweed (*Polygonum amphibium*) and buckbean (*Menyanthes trifoliata*); and along the wet shoreline, mudwort (*Limosella acaulis*) and marsh gnaphalium (*Gnaphalium palustre*). On pockets of moist soil that rise above the shallow water table grow purple cinquefoil (*Potentilla palustris*), ladies tresses (*Spiranthes romanzoffiana*), primrose monkeyflower (*Mimulus primuloides*), and, amidst the mossy terrain, the tiny sundew plant.

Yellow pond-lilies occupy the last vestiges of open water in a lake transitioning from swamp to meadow to forest.

Most classification schemes include **montane meadow** as a recognizable yet generalized community that includes wet to dry meadows ranging from the upper montane high into the subalpine zone. Wet meadows with standing water typically are composed of moisture-loving members of the grass, rush, and sedge families, often to the exclusion of other herbaceous perennials. As surface water recedes, plants tolerant of waterlogged soils begin to appear, such as great polemonium (*Polemonium occidentale*), swamp onion (*Allium validum*), bog saxifrage (*Saxifraga oregana*), toad lily (*Montia chamissoi*), rein orchid (*Platanthera leucostachys*), sticky willow-herb (*Epilobium ciliatum*), long-stalked starwort (*Stellaria longipes*), and bistort (*Polygonum bistortoides*). Some meadow habitats in the Basin contain fine, impermeable soils that collect water in spring and subsequently dry to a clay hardpan by mid-summer. This harsh **vernal pool** environment is left to annuals such as the showy porterella (*Porterella carnosula*), woolly marbles (*Psilocarphus brevissimus*), and needle navarretia (*Navarretia intertexta* ssp. *propinqua*). In better drained, rocky-bottomed meadows, such as those at east Sagehen Meadows, the late spring to early summer snowmelt kicks off a series of wildflower blooms, which in rough sequential order may include buttercups, rock star (*Lithophragma glabrum*), camas lily (*Camassia leichtlinii*), Sierra lewisia (*Lewisia nevadensis*), elephant's head (*Pedicularis groenlandica*), Bridge's gilia (*Gilia leptalea*), meadow penstemon (*Penstemon rydbergii*), yampah (*Perideridia* spp.), white brodiaea (*Triteleia hyacinthina*), bog mallow (*Sidalcea oregana*), alpine aster (*Aster alpigenus*), western aster (*Aster occidentalis*), and, occasionally, hiker's gentian (*Gentianopsis simplex*).

Subalpine Life Zone

The subalpine zone begins somewhere near 8,000 feet elevation and continues up to approximately 9,000 to 10,000 feet, occupying the elevation range between the dense fir forests and wooded stream channels of the lower Basin and timberline. This region is characterized by open terrain checkered with conifer groves, steep talus slopes, seep meadows, and rocky snowmelt streams. The subalpine zone is best developed along the western crest of the Basin, through Castle Peak head-ing south along the Pacific Crest Trail via Tinkers Knob and Twin Peaks into the Desolation Wilderness and beyond to Carson Pass. Mount Rose and the Freel Peak-Jobs Sister complex also have significant subalpine terrain.

The subalpine zone in the Basin is characterized by cooler temperatures, less vegetation, and an often extensive history of glaciation. As a consequence, soils in the subalpine zone typically are poorly developed, thin and rocky, with little organic matter. Deeper soils, where they occur, are often waterlogged for good portions of the shorter growing season. In some years snowpacks remain on the ground into August. One result of these varying soil and moisture conditions is that plant communities are generally less distinct here than at lower elevations. The **lodgepole pine forest**, a well developed forest community in the southern and central Sierra, is uncommon in the Basin, although it may occur occasionally on dry, higher-elevation slopes such as Mount Rose or Freel Peak. Lodgepole pines are abundant, however, in mixed-species assemblages throughout Tahoe's subalpine habitat. In contrast to their role as a pioneer species at lower elevations, lodgepoles flourish as seemingly permanent fixtures in the marginal soils of the subalpine zone through their ability to tolerate alternately saturated and parched conditions within the same growing season.

In the Basin the typical **subalpine forest** community is a mixture of several conifer species. Lodgepole pine, and occasionally Sierra juniper, continue to appear in the mixed assemblage up to timberline. Jeffrey pine and red fir may occur sporadically at lower elevations, but are gradually replaced as one climbs by western white pine and mountain hemlock, the latter occurring almost exclusively on cool, moist, north- and east-facing exposures. Above 8,500 feet whitebark pine is also common, forming extensive stands on the windswept slopes of Mount Rose and Freel Peak. Many high-elevation trees grow in twisted, stunted forms (known as *krummholz*) as they ap-

proach timberline. The best known krummholz species in the
Basin is whitebark pine, but Sierra juniper, mountain hemlock, lodgepole pine, and even an occasional Jeffrey pine also
grow in this manner. The krummholz Sierra juniper populations found along the western crest of the Basin are unique in
the Sierra.

Most micro-communities in the subalpine zone are extensions of similar habitats found at lower elevations and thus
share many of the same species. Lakes and meadows are often
fringed with lodgepole pines and shrubs in the heath family
such as red heather, alpine heather (*Cassiope mertensiana*), Sierra bilberry (*Vaccinium caespitosum*), and alpine laurel (*Kal-
mia polifolia*). Seeps are a wetland community that may occur
at any elevation, but are particularly common in the high country where slowly melting snowpacks provide constant moisture to downslope terrain. In these moisture-rich environments, one may commonly find alpine shooting star (*Dodecatheon alpinum*), little elephant's head (*Pedicularis attollens*),
Maclosky's violet (*Viola macloskeyi*), corn lily (*Veratrum californicum*), alpine everlasting (*Antennaria media*), sibbaldia
(*Sibbaldia procumbens*), marsh marigold (*Caltha leptosepala*),
fan-leaf cinquefoil (*Potentilla flabellifolia*), and, infrequently,
alpine gentian (*Gentiana newberryi*). Where wet soils are thin,
above rock outcrops, annuals such as Brewer's monkeyflower
(*Mimulus breweri*) or miniature gilia (*Gilia capillaris*) may
predominate. Moist, rocky slopes reveal the high-elevation
species Sierra primrose (*Primula suffrutescens*), Drummond's
cinquefoil (*Potentilla drummondii*), rock fringe (*Epilobium
obcordatum*), Drummond's anemone (*Anemone drummondii*),
and seep-spring arnica (*Arnica longifolia*). Small cascading
streams add grass of Parnassus (*Parnassia* spp.), peak saxifrage
(*Saxifraga nidifica*), and Eastwood willow (*Salix eastwoodiae*)
to the wetland mix.

In drier environments, among rock outcroppings, shrubs

such as red elderberry (*Sambucus racemosa*), Alpine prickly currant (*Ribes montigenum*), and shrubby cinquefoil (*Potentilla fruticosa*) associate with perennial herbs, including explorer's gentian (*Gentiana calycosa*) and Douglas's catchfly (*Silene douglasii*). Growing out of rocky terraces and cracks are granite gilia (*Leptodactylon pungens*), Lobb's buckwheat

Chaparral and forest occur amidst granitic outcroppings at Shirley Canyon.

(*Eriogonum lobbii*), and pink alum root (*Heuchera rubescens*). On the open slopes one finds mountain flax (*Linum lewisii*), lavender gilia (*Ipomopsis tenuituba*), Shasta knotweed (*Polygonum shastense*), marum-leaved buckwheat (*Eriogonum marifolium*), raillardella (*Raillardella* spp.), Tahoe lupine (*Lupinus argenteus* var. *meionanthus*), whorled penstemon (*Penstemon heterodoxus*), and woolly sunflower.

Alpine Life Zone

Distinct alpine plant communities are best observed in Tahoe on the windswept slopes and summits of Mount Rose and the Freel Peak-Jobs Sister complex. These communities are descendants of tundra species to the north and desert species to the east that have evolved in isolation from lower-elevation flora during glacial periods when the intervening subalpine terrain was covered for long periods by snow and ice. For simplicity, the alpine zone will be defined here as the region above timberline, where trees do not grow. In Tahoe, timberline ranges from under 9,000 feet for windswept ridges along

Subalpine wildflower gardens are abundant below Pyramid Peak.

the Sierra crest to almost 10,000 feet on the highest mountains in the Basin. With recession of the glaciers, the clear separation between alpine and lower-elevation communities has
diminished somewhat; many Tahoe communities above timberline resemble extensions of subalpine vegetation more than
distinct floras. Other high-elevation outposts such as Castle
Peak, Slide Mountain, Tinker's Knob, Mount Tallac, Dick's
Peak, Ralston Peak, or Roundtop offer noticeably different plant
assemblages.

However distinct the community makeup may be, all plants
growing above timberline confront a similar range of highly
stressful environmental conditions. As a general rule, the upper elevation range for trees is a function of temperature; as
average daytime temperatures in summer decrease with increasing elevation, trees are unable to photosynthesize sufficient carbohydrates to support woody growth. Topography
clearly also plays a role in upward distribution by affecting wind
velocities and snowpack depths. Winds limit photosynthetic

activity, both by decreasing temperature and by increasing evapotranspiration, which leads to loss of moisture. Moisture stress is exacerbated on summits and steep south-facing exposures where windblown snowpacks are shallow and usually gone by early summer, having percolated into the rocky, well drained alpine soils. As a result, most alpine plants are perennials that grow quickly and flower early in the season, typically before plants in the subalpine zone, thus avoiding midsummer drought.

Wetland environments in the alpine zone are less common than at lower elevations and usually take the form of seeps or narrow streams below melting snowbanks. Many moisture-loving subalpine species range above timberline, some of which, such as alpine saxifrage (*Saxifraga tolmiei*) or alpine buttercup (*Ranunculus eschscholtzii*), actually seem to prefer these high elevations. On open gravelly slopes one may find showy penstemon, pussypaws, daggerpod (*Phoenicaulis cheiranthoides*), or pink star onion (*Allium platycaule*). Amidst rock outcroppings are shrubs such as white-stemmed golden-bush (*Ericameria discoidea*), singlehead goldenbush (*Ericameria suffruticosa*), alpine gooseberry (*Ribes lasianthum*), and wax currant (*Ribes cereum*), the latter three having ranges that extend down into the upper montane zone. Perennial herbs include shining daisy (*Erigeron barbellulatus*), shaggy hawkweed (*Hieracium horridum*), Sargent's catchfly (*Silene sargentii*), alpine paintbrush (*Castilleja nana*), and Gray's bedstraw (*Galium grayii*). On higher peaks one may also find mountain sorrel (*Oxyria digyna*), rose buckwheat (*Eriogonum rosense*), woolly groundsel (*Senecio canus*), and alpine gold (*Hulsea algida*).

The summits of the highest ridges and peaks in the Basin are particularly harsh environments, with little snowpack, constant winds, sparse soil, and cold temperatures. Plants living under such conditions grow close to the ground, in mat or cushion form, or in basal rosettes. Two common mat-forming shrubs occurring at these elevations are short-stemmed steno-tus (*Stenotus acaulis*) and low sagebrush (*Artemisia arbuscula*). Plants in this environment are perennials that survive by maintaining most of their growth below ground. This allows them to access elusive moisture sources and protect exposed parts from the extreme conditions above. Such species include cut-leaved daisy (*Erigeron compositus*), dwarf alpine daisy (*Erigeron pygmaeus*), Davidson's penstemon (*Penstemon davidsonii*), globe gilia (*Ipomopsis congesta* ssp. *montana*), Alpine cryptantha (*Cryptantha humilis*), showy polemonium (*Polemonium pulcherrimum*), locoweeds (*Astragalus* spp.), hairy paintbrush (*Castilleja pilosa*), drabas (*Draba* spp.), cushion phlox (*Phlox condensata*), and butterballs (*Eriogonum ovalifolium*). Most of these species belong to genera common in drier habitats to the east, relict plants that have exploited their ancestral desert adaptations to survive in the extreme alpine environment.

East-Side Communities

The Tahoe Basin has no true examples of the distinct vegetative communities of the eastern Sierra, **pinyon-juniper woodland** and **sagebrush scrub**. However, many important associates of these communities occur in the Basin, becoming more common toward the east or in pockets on the north and south where precipitation is noticeably lower than on the western side. Along the east side of the Carson Range, generally above 6,000 feet, Jeffrey pines form open, almost monocultural forests with an understory of bitterbrush, rabbitbrush (*Chrysothamnus nauseous*), and mountain sagebrush. As one moves downslope, out of the Basin, annual precipitation drops below twenty-five inches and forested Jeffrey pine habitat gives way to sagebrush scrub. Two dry-adapted perennials common in this community are prickly poppy (*Argemone munita*) and blazing star (*Mentzelia laevicaulis*). Both species have large, showy, light-colored flowers that are easily visible to nighttime moth pollinators.

Pinyon-juniper woodland is absent from the Basin, beginning approximately twenty miles southeast of Monitor Pass. Isolated patches of pinyon pine (*Pinus monophylla*), an abundant conifer in this community, can be found on the eastern flanks of Mount Rose or along the drier slopes of mountains rising out of Hope Valley to the south. A similar distribution pattern is seen for curl-leaf mountain mahogany (*Cercocarpus ledifolius*), which occupies dry hillsides in the northeastern part of the Basin and can also be found in abundance along south-facing slopes in the vicinity of Freel Peak and Carson Pass.

HISTORY OF THE TAHOE BASIN

For thousands of years prior to the arrival of European settlers, the Tahoe Basin was occupied by the Washoe Indians, who migrated from winter camps on the eastern side of the Carson Range into the Basin to replenish food supplies in spring and summer. The name Tahoe is derived from a European mispronunciation of the Washoe term for edge of the lake, where the different migrating groups came together during the summer months. In early summer the Washoe fished the spawning runs of native cutthroat trout in streams entering the lake. At the end of the spawning season, Washoe families turned to gathering berries, roots, vegetables, spices, and medicinal products from the local flora of the Basin, while the hunters returned to the east side to pursue game, particularly deer and rabbits. In early fall the Washoe gathered for the annual nut harvest from the pinyon pines growing in the southeastern sections of the Carson Range.

The first settlers to record a sighting of Lake Tahoe were members of Captain John C. Fremont's party (including their guide, Kit Carson), who described the lake from a vantage point on or around Red Lake Peak in February 1844. Later that same year a small band in the north split off from a group of emigrating families and followed the lower Truckee River south to the shores of the lake. With the advent of the Gold Rush in 1848, interest intensified in establishing routes from the eastern Sierra to the western slope. In 1852 John Calhoun Johnson pioneered a route from the southern part of Lake Tahoe to Placerville over what is today known as Echo Pass. Five years later this route was connected by stagecoach to Carson Valley over Luther Pass to the south. During the 1860s a wagon road was constructed over Donner Pass, followed in 1868 by a railroad line running from San Francisco through what is now Reno, Nevada.

The discovery of the Comstock lode in Virginia City, Nevada, in 1859 attracted yet more visitors to Tahoe. By the mid-1870s roads, rail lines, and steamships provided access for travelers to destinations throughout the Basin. Commercial enterprises sprang up to meet the needs of the silver mining communities to the east. Fishing boats plied the waters of the lake. Meadowlands were converted to livestock and dairy operations. In response to unremitting demands for building materials and firewood, intensive logging began on the eastern shores of the lake, gradually spreading throughout the Basin as the most accessible timber was cut down and hauled over Spooner Summit to Carson City.

By the turn of the century, the resources of the Basin and its surrounding areas were largely depleted. Native cutthroat trout had been fished out and replaced by non-native species. (Commercial fishing was outlawed in 1917.) Forests surrounding the Basin had been cut down and the land further scarred through overgrazing by sheep and rampant forest fires. The collapse of the mining industry in Virginia City by the late 1890s reduced overall demand for lumber and food commodities, constricting the economic base for many communities

within the Basin and thus setting the stage for the gradual shift
to tourism as Tahoe's main industry in the twentieth century.

The rise of tourism was marked by various events in Tahoe's
history: the launching of the fancy steamship *Tahoe* by former
lumberman and local entrepreneur Duane Bliss in 1896; the
opening of the luxury Tahoe Tavern hotel just south of Tahoe
City in 1901; and the completion of a road around Emerald
Bay in 1913, connecting the western access routes running over
Donner and Echo summits. Private lots around the lake, previously owned by lumber and railroad interests, were now sold
to real estate developers for the construction of summer homes,
the most elegant of which is the Vikingsholm Castle built by a
Chicago widow in 1929 at Emerald Bay. Tourist development
created a new economy in the Basin, and settlements (such as
the one at South Lake Tahoe) were soon established to provide housing for the growing number of permanent residents
serving the tourist clientele. Winter activities in Tahoe began
to pick up after World War II with the opening of ski resorts,
including Squaw Valley in 1955 and Heavenly Valley in 1956.
The selection of Squaw Valley as the site of the 1960 Winter
Olympics forever marked the Basin as a world-class destina-
tion for winter sports, thus providing a two-season tourist industry and an increased dollar base for local economies. Fuel-
ing continued growth was the rapid postwar rise of the gam-
bling industry on the Nevada side of the lake and in Reno,
thirty miles to the northeast.

Land Ownership

During the nineteenth century large sections of land within
the Basin were given away by Congress to individuals and large
corporations, through laws such as the Homestead Act or the
Railroad Acts of the 1860s and 1870s. The story of Tahoe's land
ownership in the twentieth century is mostly about efforts by

federal and state governments to reacquire some of that same land for public use and enjoyment. In 1899 Congress established the Lake Tahoe Forest Reserve, a largely uninhabited area in the southwest portion of the Basin, which included the lands around Glen Alpine, the shoreline between Rubicon Bay and Camp Richardson, and the large tract of granite high country today protected as the Desolation Wilderness, the most heavily visited wilderness area in the nation. The forest reserve was extended in 1905 to include most of the still public land on the California side of the Basin, mostly on higher slopes around the lake.

Tahoe's scenic attractions generated a series of proposals in the early part of the century to turn Lake Tahoe into a national park. These were opposed by timber, mining, and grazing interests, as well as foothill residents who feared the loss of economic opportunities west of the Basin. Ultimately the idea was dropped, based on the Park Service's conclusion that extensive development around the lake had already made it impossible to include Tahoe's natural beauty within the national park system. Instead, federal and state governments focused on acquiring available private land within the Basin. The Forest Service was able to purchase more than 3,000 acres in 1950, leading to the eventual creation of the Pope and Baldwin beach areas and the Forest Service's south shore visitor center. Meeks Bay and the inland property leading into the Desolation Wilderness were obtained in the 1970s. Other large parcels in the north and south, which could not be acquired, were subsequently developed into Incline Village and the south shore Tahoe Keys Marina.

State purchases of land mirrored these federal efforts. In combination with other federal land purchases, Nevada's acquisition through eminent domain of a large private landholding created an unbroken stretch of public ownership from the lake shore up into the Carson Range, from Incline in the north to Zephyr Cove in the south. California was also busy, purchasing what is now the Tahoe State Recreation Area (just east of Tahoe City) in 1927 and Bliss Beach in 1928. The state obtained the area around Emerald Bay, including the Vikingsholm Castle, in the 1950s and Sugar Pine Point in 1964. California's most recent major purchase, in 1978, was of Burton Creek, an area just northeast of Tahoe City that includes the wildflower-rich Antone Meadows. In 1984 the California legislature created the Tahoe Conservancy, a state agency empowered to purchase environmentally sensitive parcels. Since 1985 the Conservancy has purchased more than 4,500 parcels totaling almost 6,000 acres. In addition, the Conservancy has authorized expenditures for numerous projects related to erosion control, public access, wildlife enhancement, and resource restoration.

Environmental Degradation and Regulation

Ever-increasing residential and commercial growth in the Basin has led to a number of environmental problems, including air pollution (mostly from car exhaust), destruction of sensitive habitat, and increased erosion and siltation. During the 1960s and 1970s sewage leakage into the lake and accompanying algae blooms threatened the lake's famous water clarity. The nonprofit League to Save Lake Tahoe was created in 1965 in order to inject an environmental voice into the debate over Tahoe's future. Several governmental organizations were formed to address the regional issues that increasingly confronted planners. By the late 1960s both California and Nevada had created state planning agencies to deal with growth in the Basin, often facing staunch criticism and legal action from local commercial interests.

The need for region-wide planning led to the federal passage in 1969 of a bi-state compact that created the Tahoe Re-

gional Planning Agency (TRPA). The relative inability of the
TRPA to control accelerating development during the 1970s
resulted in a 1980 compact that gave the TRPA greater regula-
tory powers. Amendments to the federal Clean Water Act also
bolstered the TRPA's authority to control non-point source
pollution. Under these authorities the TRPA must adopt five-
year regional plans designed to meet environmental threshold
carrying capacities for water and air quality, soil conservation,
and vegetation and wildlife resources. Over the last decade regional plans and their accompanying threshold standards have
been contested in court, both by environmental groups such
as the League to Save Lake Tahoe and by the more recently
formed property owners' group, the Tahoe-Sierra Preservation
Council.

The TRPA's 1996 regional plan for the Tahoe Basin states
that the Basin still fails to meet federal and state air pollution
standards for ozone and visual range. Diminished water quality, particularly from the overloading of nutrients from runoff
and air particulates, remains a serious problem. Logging and
years of fire suppression have left the Basin with crowded, even-
aged stands of forest that lack normal species diversity. Growth
continues to threaten the natural attributes and tranquillity of
the area. In recent years the alarming disappearance of the once
common mountain yellow-legged frog (*Rana muscosa*) suggests that airborne pollution from the Central Valley and ur-
ban centers to the west may pose long-term threats to Tahoe's
environmental health. In the coming years government officials will address these issues with the input of interested groups
and individuals.

Native Plants and Conservation

Because of its unique geographical history and location,
California is home to an extraordinary number of native plant

species, by far the largest state flora in the nation. Over one-third of its native taxa are endemic to the state, occurring nowhere else in the world. While some of these endemics (such as the giant sequoia or coastal cypresses) are relict species with an ever-decreasing range, most are still in the process of evolutionary diversification. This is particularly true of genera such as *Eriogonum, Clarkia, Lupinus, Arctostaphylos,* and *Astragalus,* not to mention the numerous radiating taxa in the sunflower family. The conservation of native plants ensures that this long-term evolutionary development can continue to occur. California's natives are increasingly threatened by habitat alteration, pollution, recreational activities, and the introduction of non-native plants and animals. Non-native animals such as wild pigs or cattle may disturb habitat or feed on native plants that lack the defenses a species might normally develop through coevolution with a native predator. Non-native plants, freed from the constraints of their natural environment (insects, pathogens, herbivores, competing species) and usually better adapted to altered habitats, often outcompete native plant species for essential resources, leading to a depletion of native populations. Unlike many regions in the state, however, Tahoe's seasonally harsh climate ensures that native plants, with their more specific adaptations, form the vast majority of the local flora.

The California Native Plant Society (CNPS) is a nonprofit organization dedicated to enhancing the understanding and appreciation of California's native plants. CNPS seeks to preserve native plants in their natural habitats through research, education, and conservation. CNPS was founded in 1965 and now has more than 10,000 members in thirty regional chapters throughout the state. The Tahoe Chapter of CNPS was formed in 1983 and offers regular native wildflower hikes during the summer. The Northern Nevada Native Plant Society also offers occasional summer hikes in the Basin.

The snowpack still lingers into early summer in a red fir forest community.

PLANT
DESCRIPTIONS

SEEDLESS VASCULAR PLANTS (FERNS AND HORSETAILS)

Seedless vascular plants reproduce by releasing *spores*, which germinate, forming small *haploid* (having one chromosome of each chromosome type) structures known as *gametophytes*. The photosynthetic gametophyte produces sperm and eggs. Sexual reproduction occurs when sperm move through films of moisture to reach an egg on another gametophyte. After fertilization the mature *diploid* (having two chromosomes of each chromosome type) plant germinates and begins to grow. Spores are produced by a specialized organ called a *sporangium*, which in ferns typically occurs on the underside of the leaves. The many *sporangia* may cluster in a *sorus* (plural *sori*), often protected by a covering called an *indusium*. On horsetails the spores develop inside cone-shaped clusters, which form at the growing tips of the plant. Drawings of these structures can be found in the Glossary. Despite Tahoe's short growing season and summer-drought climate, the Basin flora includes a variety of fern and horsetail species, many of which survive in rocky, open, and seemingly dry terrain.

BRACKEN FAMILY (DENNSTAEDTIACEAE)

This mostly tropical family has around seventeen genera with over 300 species. *Pteridium* is sometimes included in its own family. Bracken fern is Tahoe's only representative of this family.

Bracken fern (*Pteridium aqualinum* var. *pubescens*)

- **2 to 4 feet**
- **mid-season**
- **semi-shady, moist to dry forest**

Bracken fern is common at lower elevations in mixed coniferous forest. The large *leaf blades* are widely triangular with significantly reduced upper leaflets. The non-clustered *sporangia* occur along the outer margins of the underside leaflet lobes. Although the young fronds of bracken fern may be eaten, the mature plant is toxic in large quantities. **Leather grapefern (*Botrychium multifidum*)** occurs in the southwest Basin in moist, semi-shady habitats up to 9,500 feet. This species has large, fleshy, infertile leaves with rounded leaflets, which grow close to the damp terrain. The dissimilar fertile leaves grow erect, up to two feet tall. **Yosemite moonwort** (*Botrychium simplex*) is similar, but much smaller, with four-inch-long, fan-shaped infertile leaf blades. This uncommon species has been collected in moist habitat at Haypress Meadows.

WOOD FERN FAMILY (DRYOPTERIDACEAE)

This large, mostly tropical family includes lady ferns (*Athyrium*), wood ferns (*Dryopteris*), sword ferns (*Polystichum*), cliff ferns (*Woodsia*), and fragile ferns (*Cystopteris*).

Alpine lady fern (*Athyrium alpestre* var. *americanum*)

- **1 to 2 feet**
- **mid-season**
- **moist, rocky habitats**

This species is common in the upper montane and lower reaches of the subalpine zone. It can be distinguished from bracken fern by its smaller size, lanceolate leaf blades with gradually tapered leaflets, and the distinct rounded *sori* that dot the underside of the leaflet margins. **Lady fern** (*Athyrium filix-*

femina var. *cyclosorum*) occurs mainly in the southwest Basin. The *sori* are oblong or j-shaped, occur near the center of the leaflet undersurface, and are covered by a similarly shaped *indusium*. **Fragile fern** (*Cystopteris fragilis*) is common in the Glen Alpine region and occasional elsewhere on moist, rocky slopes. It can be recognized by its more widely spaced leaflets and hood-like *indusium*.

Holly fern (*Polystichum lonchitis*)

- **1 to 2 feet**
- **mid-season**
- **moist, rocky crevices**

This widespread species is rare in the Tahoe Basin, occurring in the north and along the Eagle Falls trail to upper Velma Lake. The once-pinnate, leathery leaves (consisting of the leaf *petiole* and the leaf *blade*) range from four to twenty-four inches long. The leaflets are slightly serrated, with the first lobe noticeably larger than the rest. **Kruckeberg's sword fern** (*Polystichum kruckebergii*) has been found in dry crevices in the Echo Lakes region. It may be distinguished by its shorter leaves and *petiole* (about one-tenth the size of the leaf *blade*), and more deeply lobed lower leaflets. The rare **cliff fern** (*Woodsia scopulina*) occurs in rock crevices and on steep slopes near Glen Alpine and Ralston Peak.

Bracken fern

Alpine lady fern

Holly fern

The *indusium* of this genus holds the *sorus* in a cup-like structure, from below.

HORSETAIL FAMILY (EQUISETACEAE)

True relics of the Paleozoic, horsetails have survived for approximately 380 million years. The horsetail family has a single genus, *Equisetum*, and fifteen or more species worldwide, five of which occur in California. Horsetails are referred to as scouring rush, a dual reference to their similar appearance to members of the non-related angiosperm rush family (Juncaceae) and to the relatively high silica content of their outer cellular tissue, which native Americans used as fine sandpaper and early settlers used in washing cookware.

Common horsetail (*Equisetum arvense*)

- **4 to 24 inches**
- **early season**
- **wet to moist environments**

Common horsetail occurs throughout the Basin below 7,000 feet, sprouting with both fertile and infertile stems from rhizomes. The infertile stems grow erect with whorled linear branches resembling a horse's tail. The inconspicuous scale-like leaves are whorled and fused into a toothed sheath around each stem node. The fertile stems are shorter, often yellowish brown, and unbranched, with six to fourteen teeth per nodal sheath. The *sporangia* occur in the terminal cone atop the fertile stem. Generally uncommon in Tahoe, **giant scouring rush** (*Equisetum hyemale* ssp. *affine*) can be abundant on the shaded margins of low-elevation streams. The two- to seven-foot tall, unbranched stems are all fertile. There is a dark band above and below each nodal sheath, which has from twenty-two to fifty teeth.

BRAKE FAMILY (PTERIDACEAE)

The brake family has approximately forty genera and 500 species worldwide. It is well represented in the Basin by a number of dry-adapted species.

Indian dream (*Aspidotus densa*)

- **4 to 10 inches**
- **early season**
- **moist to dry rock crevices, slopes**

Relatively common from low elevations into the subalpine zone, Indian dream has small, leathery, pinnately dissected leaves with linear segments and a dark, shiny leaf stem. The indusia are found on the leaflet undersides, along the segment mid-veins. The fertile and infertile leaves of Indian dream are slightly dissimilar. The rare **five-finger fern** (*Adiantum aleuticum*) occurs in wet, rocky crevices between 7,000 to 7,500 feet in the Glen Alpine-Angora Lakes area, and together with holly fern, along the trail from Eagle Lake to Velma Lake. This delicate species can be easily recognized by its palmate leaf structure and reddish brown to shiny black petioles.

Lace fern (*Cheilanthes gracillima*)

- **2 to 6 inches**
- **early season**
- **moist to dry rock crevices**

This species is common in rocky habitats up to 9,000 feet. Lace fern has small lanceolate, twice-pinnate leaves, which are dark green with small, rounded leaflet segments. The segment undersides are covered with light colored scales, which resemble dense woolly patches. The

Common horsetail

Indian dream

American parsley fern

Lace fern

golden with age, giving lace fern a colorful autumn appearance in the middle of summer.

American parsley fern
(*Cryptogramma acrostichoides*)

• 4 to 10 inches
• mid-season
• moist to dry rock crevices, slopes

sporangia are often obscured by the scales and the recurved margins. Old leaves turn rust-colored and

American parsley fern is distinctive for its two noticeably different types of leaves. The narrower fertile leaves stand erect, presumably aiding in spore dispersal. Their linear leaflets have rolled margins that protect the numerous brownish spores, which are released in mid-season. The wider sterile leaves have rounded leaflet lobes and sprawl out at ground level. This species is common throughout the Basin from low elevations to the summits of the highest peaks.

Bridge's cliff brake (*Pellaea bridgesii*)

- **4 to 12 inches**
- **mid-season**
- **rocky slopes, ledges, crevices**

A common fern in the Basin amidst granite outcroppings, Bridge's cliff brake can be easily recognized by its bluish green, once-pinnate leaves with rounded, thick leaflets that are unlobed and often folded slightly lengthwise. Brewer's cliff brake (*Pellaea breweri*) is more common in the southern part of the Basin. It can be distinguished by its two- to three-lobed leaflets and slightly greener, softer leaves.

NON-FLOWERING SEED PLANTS (GYMNOSPERMS)

Of the four taxonomic divisions of gymnosperms, only the Coniferophyta, or conifers, occur in northern temperate regions. Conifers in the Basin are evergreen with thick needle- or scale-like leaves and separate male and female cones. Mature male cones release sperm clustered within protective cellular capsules known as *pollen grains*, which disperse over a wide area on wind currents. Pollination occurs when a pollen grain fortuitously lands on the female reproductive parts located below the large bracts of the female cone. As a result of pollination, a second cell in the pollen grain develops into a pollen tube, which grows downward to the female egg. If fertilization occurs, the plant embryo is formed inside a hardened protective seed coat, which also encloses leftover female reproductive material as a food reserve for the young seedling.

CYPRESS FAMILY (CUPRESSACEAE)

This family is characterized by evergreen trees or shrubs with opposite, four-ranked or whorled, scale-like leaves and fleshy seed cones, which usually become hard at maturity. There are seventeen genera and 120 species worldwide. California's four genera include incense-cedar (*Calocedrus*), cypress (*Cupressus*), juniper (*Juniperus*), and cedar (*Thuja*). Tahoe's species include incense-cedar and two junipers.

Bridge's cliff brake

Incense-cedar (*Calocedrus decurrens*)

• **60 to 150 feet**
• **late season**
• **white fir forest, open slopes**

This attractive tree is common in the Basin at elevations below 7,000 feet. It is an important associate in the white fir forests that border the western shore of the lake. Its cinnamon-red bark is often mistaken for redwood or giant sequoia, neither of which occurs in the Basin. The flattened stem branches have opposite, overlapping, scale-like leaves in sets of four. The small, oblong, yellowish pollen cones grow inconspicuously at the ends of branches, while up the stems the flat, woody seed cones hang in three sets of overlapping pairs.

Incense-cedar

Sierra juniper (*Juniperus occidentalis* var. *australis*)

• **1 to 50 feet**
• **late season**
• **open forest, rocky slopes**

Sierra junipers are the oldest plants in Tahoe, with many specimens exceeding 1,000 years in age. Common in the Basin up to 10,000 feet, specimens along Tahoe's western Sierra crest grow in "krummholzed" form, the only known population to do so. The peeling bark is reddish brown, and the scale-like, bright green leaves grow in whorls of three around the stem. Sierra junipers are dioecious, with separate male and female plants. The soft, bluish seed cones of the female trees are sought by many animal species, which play an important role in seed dispersal. **Common juniper (*Juniperus communis*)** occurs on thin granitic soils in forested openings, along lake edges, and on open slopes to over 9,000 feet. This circumboreal species grows as a prostrate shrub, with blue-green foliage and needle- to awl-like leaves.

PINE FAMILY (PINACEAE)

The pine family consists of large, usually evergreen, needled trees with unisexual reproductive structures (cones) occurring on the same tree (a condition known as *monoecious*). The male pollen cones are non-woody and deciduous. The female seed cones (usually referred to as pine cones) are woody with bracts and persistent scales. The seeds are borne in pairs, on the upper side of each scale. Pines, firs, spruces, Douglas-fir, and hemlocks are all in this family, contributing well over half the world's supply of commercial timber.

White fir (*Abies concolor*)

- **50 to 190 feet**
- **late season**
- **semi-moist to dry, cool habitats**

White fir is common below 7,500 feet, where it forms its own mixed forest community. It has a rounded crown and smooth, white-gray bark that ages to a deeply furrowed dark brown on mature trees. Like all firs, it has single, unbundled needles that leave a flat scar on the twig after they fall. The seed cones grow in a single season and sit erect on upper branches. The cones generally disintegrate on the tree, scattering seeds about the forest understory and leaving behind the upright, naked cone axes to greet the approaching winter season. Fir seeds are a favorite food of Douglas squirrels, which may leave dismembered—or occasionally intact—fir cones on the forest floor as a reminder of their presence high in the canopy above.

Red fir (*Abies magnifica* var. *magnifica*)

- **60 to 180 feet**
- **late season**
- **semi-moist, cool habitats**

Red fir is abundant on cool exposures in deep soils from 7,000 up to 9,000 feet, forming almost monocultural forests in preferred habitats. Mature red fir can be distinguished from white fir where

Sierra juniper

White fir

Red fir

needles. White fir needles are longer, flattened, and twisted 180 degrees at the base. In addition, red fir cones may be over eight inches long, almost twice the length of those of white fir. Red firs are particularly vulnerable to lightning strikes, which often act as agents of secondary succession by clearing openings in the red fir forest canopy.

their ranges overlap by its deeply furrowed, purplish red bark and tightly needled, jigsaw-puzzle branches, which project out horizontally from the main trunk. Younger red firs, with grayish bark, can be distinguished by their smaller, untwisted, rounded

Whitebark pine (*Pinus albicaulis*)

• 1 to 60 feet
• mid- to late season
• high-elevation slopes, summits

Pines are distinguishable from firs and hemlocks by their variable number of needles, which are bundled together in a bracted sheath that may persist for many growing seasons. Whitebark pine is a true subalpine species, rarely occurring below 8,000 feet, but forming dense, shrubby "krumm-holzed" stands near the summits of high peaks in the Basin such as Mount Rose or Freel Peak, typically bent at the level of the snowpack, sometimes less than a foot above the ground. The bark is gray-white, and the five thick needles are dark green, stiff, and clustered together. The small, ovate seed cones are brownish purple and disintegrate slowly on the tree's upper branches if they are not first eaten by squirrels, Clark's nutcrackers, or other subalpine fauna.

Lodgepole pine (*Pinus contorta* ssp. *murrayana*)

• 1 to 120 feet
• mid- to late season
• diverse habitats

Lodgepole pine is a true generalist, able to survive and prosper over a wide range of elevations and habitat types. It is most abundant in moist areas along lake and meadow edges, aided by a root system that, unlike that of other conifers, is able to tolerate water-logged, anaerobic environments. It is also common on the thin, rocky soils that overlay granitic bedrock in higher-elevation watersheds, flourishing in these alternately soggy to dry soils by controlling rates of water uptake and loss from transpiration. Lodgepole has two-needled bundles, cornflake-like bark, and small, spherical, pointed cones, which open en masse following fire. *Contorta* refers to the tree's ability to grow prostrate, or twisted around large boulders on higher, windy exposures. The common name comes from native Americans' use of the young trees as poles for their teepees.

Jeffrey pine (*Pinus jeffreyi*)

• 2 to 160 feet
• late season
• dry, open to semi-open habitats

Jeffrey pine is common on drier soils from lake level to around 8,000 feet, with individual trees occurring well into the subalpine zone. At lower elevations on exposed rocky slopes and in the drier eastern portion of the Basin, it is the most abundant conifer. It is recognizable by its three long, grayish blue needles and reddish brown, deep-furrowed bark (which exudes a dis-

Lodgepole pine

Jeffrey pine

Whitebark pine

tinctly vanilla-like odor on warm days) and large oblong seed cones with recurved scale tips, which are gentle to the touch. The similar **ponderosa pine** (*Pinus ponderosa*) is occasional in Tahoe below 6,500 feet, where it is known to hybridize with its higher-elevation relative. Ponderosa pine has shallowly furrowed, odorless bark, and small ovoid cones with spreading, prickly scale tips. Both species are highly drought-resistant because of their deep root systems and ability to photosynthesize under moisture stress. The uncommon **Washoe pine** (*P. washoensis*), distinguishable by its shorter needles and seed cones, has been found on the east side of Mount Rose in the Galena Creek watershed.

Sugar pine (*Pinus lambertiana*)

• 90 to 220 feet
• late season
• semi-open, moist to dry
  habitats

The largest pine in the Sierra, sugar pine was logged extensively in the late nineteenth century, but is still occasional at low elevations around the lake. It is best known for its long banana-shaped cones, which grow up to eighteen inches long and hang down from long, horizontal branches. The sweet sap was eaten by Native Americans. Sugar pine is a member of the five-needled group known as "white pines" for the generally lighter color of their bark. The higher-elevation **western white pine (*Pinus monticola*)** occurs from 7,000 to 9,000 feet, most often with lodgepole pine and mountain hemlock in the subalpine forest. Like sugar pine, its five-needled bundles appear light and airy. It is distinguished by its upper branches, which turn upward towards the tip, and by its similarly shaped but smaller seed cones. The Great Basin species, **single-leaf pinyon pine (*P. monophylla*)**, occurs sporadically on the east slopes of Mount Rose and the eastern peaks above Hope Valley. The large seeds, known as pinyon nuts, were a dietary staple of the local Washoe.

Mountain hemlock (*Tsuga mertensiana*)

• 2 to 100 feet
• late season
• semi-moist, north- and east-
  facing slopes

Mountain hemlock is common on cool exposures from 7,500 to 9,700 feet, gradually replacing red fir at higher elevations. It is recognizable by its drooping top, which gives the hemlock forest a Tolkienesque quality. At higher elevations on exposed terrain, it grows prostrate. The narrow-ridged bark is purple-brown, and the two- to four-inch oblong seed cones hang at the ends of nodding branches. The single needles are short, rounded in cross section, and often glaucous blue-green. In contrast to the firs, hemlock needles leave behind a distinctly raised, woody base on the twig when they are removed. The closely related Douglas-fir (*Pseudotsuga menziesii*) does not naturally occur in the Basin, although planted trees may occasionally be found growing at lower elevations around the lake.

FLOWERING PLANTS (ANGIOSPERMS)

With a few exceptions, most angiosperms in the Basin are either woody shrubs or herbaceous annuals or perennials. Plants may be winter-deciduous or evergreen,

Sugar pine

Mountain hemlock

usually in response to moisture and length of the growing season. Angiosperms are best known for their reproductive organs, collectively referred to as the flower. Flowers in the Basin come in many different shapes and sizes, the result of millions of years of evolutionary development and variable reproductive success.

MONOCOTS

The *monocotyledons* have one initial embryonic leaf, flower parts in sets of three, and parallel leaf venation. In some monocots the primary taproot is quickly replaced by adventitious roots growing from the stem, forming a fibrous root system. Monocots are represented here by the iris, lily, and orchid families. Other monocots growing in the Basin, such as those in the widespread grass, rush, and sedge families, and several mainly aquatic families, are not covered.

WATER-PLANTAIN FAMILY (ALISMATACEAE)

This aquatic family has approximately twelve genera and close to 100 species worldwide, mostly in the northern hemisphere. The flowers have three, usually green sepals, three petals, and six or more stamens and pistils. The fruit is a beaked achene.

Arrowhead (*Sagittaria cuneata*)

- 4 to 12 inches
- mid-season
- open, shallow standing water

This species occurs infrequently in shallow marshes and lakes such as Osgood Swamp or Spooner Lake. Arrowhead is named for the distinctive shape of its floating or emerging leaf blades. The conspicuous flowers are unisexual, the female occurring below the male flowers on the same inflorescence stalk. The male and female parts of both flower types are surrounded by three round to scallop-shaped white petals. *Sagittaria* species are often called duck or tule-potato for their root bulbs, which were a dietary staple for many Native American tribes.

IRIS FAMILY (IRIDACEAE)

This large and diverse family includes only two native genera in California, *Iris* and *Sisyrinchium*. The family can be identified by its parallel-veined leaves, which overlap each other along the stem, similar to the leaf structure of the grass family, Poaceae. Flowers typically have three petal-like sepals, three true petals, an inferior ovary, and three stamens. In *Sisyrinchium* the sepals and petals are similar, while in *Iris* the inner whorl of true petals appears quite different.

Western blue flag (*Iris missouriensis*)

- 1 to 2 feet
- mid-season
- moist to dry meadows

This widespread and striking species occurs up to 8,500 feet in the south Basin, from Meyers to Carson Pass and on the eastern flanks of Mount Rose. Each flower consists of three bluish purple sepals with elaborate white and yellow veins, spreading below three darker upright petals. The flower's veins serve as visual guides for bees and other insects to the nectar sources at the bases of the sepals. As the pollinator enters the flower, its already pollen-laden backside brushes against the protruding styles and stigmas, which thereupon spring upward. As it exits the flower, anthers deposit a new layer of pollen on the insect's back, which this time avoids the upright female parts, thus ensuring successful cross pollination. Native Americans used the ground-up roots as poison for their arrow tips.

Blue-eyed grass (*Sisyrinchium idahoense*)

- 6 to 18 inches
- early to mid-season
- moist meadows, stream banks

Blue-eyed grass has narrow, grass-like leaves and blue-violet, yellow-throated flowers that are borne atop erect to ascending stems in

Arrowhead

Western blue flag

Blue-eyed grass

moist to dry meadows and along stream margins at low elevations around the Basin. The less common **yellow-eyed grass** (*Sisyrinchium elmeri*) can be found in similar low-elevation habitats such as Osgood Swamp. The flowers have six orange-yellow sepals and petals, which offer a wide open landing pad for bee pollinators.

LILY FAMILY (LILIACEAE)

The lily family is characterized by flower parts in threes (with sepals and petals often indistinguishable) and six stamens. All are perennials, many possessing a fleshy root appendage known as a *bulb* that stores carbohydrates for early-season growth. The lilies defy easy taxonomic categorization; various studies have at times divided the group into nine or more families. The broadly defined Liliaceae has over 300 genera and 4,500 species worldwide, with approximately eighteen species representing nine genera in the Basin. Members of this large family include desert species in the genera *Yucca, Nolina,* and *Agave,* crops such as asparagus and onion, and ornamental plants such as lilies, hyacinths, and tulips.

Sierra onion (*Allium campanulatum*)

• **4 to 8 inches**
• **early to mid-season**
• **moist to dry, open habitat**

Sierra onion is abundant throughout the Basin from low elevations into the subalpine zone, often forming dense pink-purple patches on flat, rocky terrain. Like all members of the genus, it has a distinct onion odor and linear basal leaves, which often wither from the tip over the course of the growing season. The small flowers are composed of six unfused, indistinguishable petals and sepals that bloom in an umbel arrangement. The less conspicuous **dwarf onion** (*Allium obtusum*) grows flat on the ground in forest openings and on open sandy slopes in the upper montane and subalpine zones. Tahoe's two varieties differ by having white to slightly greenish flowers or, in the south, pink flowers, both with a purple-red midvein on each petal. **Mountain muilla** (*Muilla transmontana*) has similar white flowers but grows on an erect stem and lacks any onion odor. This Great Basin species has been collected near Truckee and can be

found today in dry meadows and slopes in the Hope Valley region.

Pink star onion (*Allium platycaule*)

• **2 to 8 inches**
• **early to mid-season**
• **rocky to sandy slopes**

This brightly colored species is locally common on snowmelt-fed, volcanic ridgetops (such as Ward Peak and Castle Peak) in the northern end of the Basin. Besides its isolated habitat preference, it can be easily identified by its semi-prostrate growth form and bright rose, star-lobed flowers, which crowd together like lonely beacons on the otherwise drab landscape.

Swamp onion (*Allium validum*)

• **20 to 40 inches**
• **mid- to late season**
• **wet meadows, seeps**

Swamp onion is particularly common in the south Basin from low elevations to 8,500 feet. The bright pink-rose flowers have exserted stamens. The long, flat-edged basal leaves stand erect, often as tall as the colorful, flat-topped flower heads. The eastern Sierra native, **aspen onion** (*Allium bisceptrum*), occurs in rocky, moist, grassy areas in the far north and south, typically blooming immediately following the camas lilies in early summer. This species is less than half the size of swamp onion, with a more open,

Sierra onion

Pink star onion

Swamp onion

long-stemmed inflorescence. **Paper onion** (*A. amplectens*) makes a rare appearance in the Basin in moist

granitic sands in the Shirley Canyon area. This species is recognizable by its delicate, white, papery flowers, which are borne atop erect six- to twelve-inch stems.

Leichtlin's mariposa lily (*Calochortus leichtlinii*)

- **4 to 12 inches**
- **mid-season**
- **drying, open habitat**

Calochortus is a large genus in California, representing some of the state's most striking flowers. (*Calochortus* is Greek for beautiful grass.) Each flower has three wedge-shaped petals and three, generally smaller, lanceolate sepals. The lower-elevation globe lilies, members of *Calochortus*, do not occur in Tahoe.

Leichtlin's mariposa lily is common in the montane zone, often forming broad patches on flat, rocky terrain where ephemeral moisture can be hoarded in underground bulbs, without the need for a deep root system. On the higher-elevation slopes of the Carson Range near Genoa Peak, one may find the similar *C. bruneaunis*, distinguishable by the bright, dark red chevron above the petal's yellowish base, the small red patch below, and the faint green vertical stripes on the outside of each petal. Mariposa lilies are named after the Spanish word for butterfly, in reference to their brilliantly colored flowers.

Sierra star tulip (*Calochortus minimus*)

• **2 to 5 inches**
• **early to mid-season**
• **moist to dry forest openings**

This species is locally common in the southwest, particularly in the Desolation Wilderness. The snow-white petals are hairless, smaller and more delicate than those of *Calochortus leichtlinii*. Occasional populations with pinkish petals and rounded margins are thought to be the products of hybridization with naked star tulip (*C. nudus*), which occurs further to the southwest near Wright's Lake.

Camas lily (*Camassia quamash*)

• **8 to 20 inches**
• **early season**
• **wet to moist rocky meadows**

This occasional species can be locally abundant in lower-elevation meadows throughout the Basin, occasionally forming vast swaths of purple that grace east Sagehen Meadows in late spring. The six indistinguishable purple petals and sepals are complemented by stamens with bright yellow anthers. The bulbs of camas lilies were a dietary staple of Native Americans, including the local Washoe, for thousands of years. One cause of the wars between early European settlers in Idaho and Chief Joseph's Nez Perce tribe was the destruction by livestock of the camas lily fields, particularly by pigs, which uproot the bulbs. The species name comes from the Nez Perce word for sweet. There are two recognized subspecies, *Camassia quamash* ssp. *breviflora* and *C. q.* ssp. *quamash.*

Spotted mountain bells (*Fritillaria atropurpurea*)

• **4 to 18 inches**
• **early season**
• **dry, open forest, rocky slopes**

This easy-to-miss plant is occasional in a variety of habitats, from semi-shady forest to windswept subalpine slopes. The one to five inconspicuous flowers droop downward, concealing six richly

Leichtlin's mariposa lily

Sierra star tulip

Camas lily

Spotted mountain bells

mottled, purplish brown and yellow petals, which spread wide open to reveal strongly exserted stamens with light yellow anthers. The linear leaves grow opposite, singly or in twos or threes at each stem node.

Normally slender, some high-elevation populations grow short and stout, often with varying flower coloration. At present, all forms occurring in the Basin are considered to be a single species.

Scarlet fritillary (*Fritillaria recurva*)

- 1 to 3 feet
- early season
- dry, open forest

The attractive and rare scarlet fritillary occurs in the north Basin below 7,000 feet, blooming in late spring amidst the open Jeffrey pine forest understory of serviceberry and bitterbrush. It grows erect with linear to lanceolate leaves whorled in groups of two to five on the lower stem and alternate above. The nodding, bright red flowers have rounded, recurved lobes. The lower-elevation plainleaf fawn lily (*Erythronium purpurascens*) occurs just outside the Basin on the Loch Leven Lakes trail off Highway 80, blooming early in the season with showy, six-lobed flowers, which are white with yellow bases.

Alpine lily (*Lilium parvum*)

- 2 to 6 feet
- mid- to late season
- moist, semi-shaded habitats

Alpine lily is one of the few orange-flowered plants in the Basin. The six lobes of the trumpet-shaped flowers are partly fused, with darker reddish orange spots on the inner tube. The plant can be recognized prior to blooming by its typically whorled, though sometimes alternating, elliptic leaves, characteristic of the genus. Leopard lily (*Lilium pardalinum*) does not occur in Tahoe, but can be found an hour's drive to the north in the Lakes Basin area off Highway 49.

Washington lily (*Lilium washingtonianum*)

- 2 to 8 feet
- mid-season
- open, dry forest, montane chaparral

The spectacular Washington lily is occasional on dry slopes, mostly in the northwest Basin, amidst open Jeffrey pine forest and chaparral, where it finds refuge from deer. The large white flowers have recurving lobes and a pinkish inner surface. Named after Martha Washington, this plant was considered by John Muir to be the finest of the Sierra lilies.

Bog asphodel (*Narthecium californicum*)

- 8 to 20 inches
- mid-season
- stream banks, lake margins, moist habitats

Bog asphodel has an extremely limited range in Tahoe, found only in the near vicinity of the Velma

Scarlet fritillary

Alpine lily

Washington lily

Bog asphodel

Lakes on the east side of the Desolation Wilderness between 8,000 to 8,500 feet elevation. The mostly

basal leaves are linear and grasslike. The bright yellow-green, starlike flowers, each with six indistinguishable petals and sepals, are borne on an erect to ascending raceme. The six erect stamens are thick, with woolly filaments, and the stigma is slightly three-lobed.

False Solomon's seal (*Smilacina racemosa*)

• **12 to 30 inches**
• **early to mid-season**
• **moist, semi-shady forest**

This attractive plant occurs throughout the upper montane zone. It has wide ovate to elliptic leaves, which clasp the erect stem. The small white flowers crowd together on a well developed *panicle* in groups of twenty or more near the top of the stem. **Star Solomon's seal** (*Smilacina stellata*) has narrower leaves and larger flowers in a *raceme* arrangement, each flower borne on a single stalk (*pedicel*) that grows from the main stem. Also known as slim Solomon, this species enjoys similar habitat, but may also occur on open slopes at low elevations. Both species produce spherical, red-purple berries in late summer. Solomon's seals are named after an Old World species whose root scars were thought to resemble the seal of King Solomon.

Tofieldia (*Tofieldia occidentalis* ssp. *occidentalis*)

• **8 to 30 inches**
• **mid-season**
• **wet meadows, stream banks**

Tofieldia is rare in Tahoe, found only at Sagehen Creek, Osgood Swamp, and in the eastern Desolation Wilderness. It occurs with bog asphodel along the subalpine stream feeding Upper Velma Lake. This delicate plant has linear basal leaves and a small, rounded inflorescence that blooms atop an erect, generally leafless stem. Each flower has six white petals and sepals. The six stamens have bright yellow anthers. These characters help to distinguish Tofieldia's spherical flower heads from the more tightly packed, five-lobed flowers of bistort (*Polygonum bistortoides*).

Golden stars (*Triteleia ixioides* ssp. *anilina*)

• **4 to 8 inches**
• **mid-season**
• **drying, open, sandy, rocky habitats**

Also known as pretty face, this species is common from low elevations through the subalpine zone. Like all *Triteleia*, it has one to three basal leaves and an umbel-shaped inflorescence with six-lobed flowers (identical sepals and petals). The lobes are yellow with dark purple stripes on the outside edges. *Triteleia* species differ from the

False Solomon's seal

Golden stars

Tofieldia

onions (*Allium*) in their lack of onion odor and their flowers, which are fused at the base into a funnel-like tube. **Mountain brodiaea** (*T. montana*) has been found in dry, open habitat at Sagehen Creek and near Echo Summit. This similar species differs from golden

stars in having equal-sized stamens that lack forked appendages on the outside of the anthers at the filament tips.

White brodiaea (*Triteleia hyacinthina*)

• 1 to 2 feet
• mid-season
• moist to dry meadows

This attractive species is occasional throughout the Basin below 7,500 feet. Similar to golden stars, it has a taller flower stalk and a denser inflorescence of white flowers with purplish green-bordered lobes. White brodiaea and other species of *Triteleia* were formerly placed in the more narrowly defined amaryllis family (Amaryllidaceae).

Corn lily (*Veratrum californicum* var. *californicum*)

• **3 to 6 feet**
• **mid-season**
• **wet meadows, stream banks**

This moisture-loving plant is abundant in wet montane meadows, often forming dense patches that exclude other plants. Beginning growth as narrow shoots pushing through the sometimes still snow-covered sod, corn lilies bloom with dozens of small white to greenish flowers on a branched inflorescence. The leaves of the plant are toxic and should not be eaten. Native Americans boiled the roots into a liquid contraceptive that, taken over several weeks, could induce permanent sterility.

Death camas (*Zigadenus venenosus* var. *venenosus*)

• **6 to 18 inches**
• **early season**
• **moist open, rocky slopes, seeps, meadows**

Death camas is common on the west side from low elevations into the subalpine zone, occurring abundantly in Shirley Canyon and Glen Alpine and less frequently toward Carson Pass. It has several linear leaves, which are significantly reduced above, and small white flowers that are borne in a *raceme* (each flower has its own stem). **Sand corn (*Zigadenus paniculatus*)** is common in drier, forested openings, meadow edges, and open volcanic slopes, but only in the far north and south Basin. It has a panicle-like inflorescence, in which few to many flowers are borne off branches that grow from the lower to mid-main stem. Unlike those of death camas, many of these branching flowers are either sterile or staminate (lacking female parts). All members of this genus are highly poisonous.

ORCHID FAMILY (ORCHIDACEAE)

With almost 20,000 species worldwide, the orchid family is second in size only to the sunflowers. While most orchids are tropical, the Tahoe Basin is home to eight species in seven genera. The tremendous variety in this family is a consequence of co-evolution between the orchid flower and specific pollinators. In the tropics, where plant diversity is high but numbers of any one species in a given area may be low, reproductive success often depends on a plant's ability to develop an exclusive relationship with a specific pollinator. Orchids have been exceptionally successful in this endeavor, developing along the way a complex floral structure in which the third petal, known as the lip, clearly differs from the other two petals and three sepals in size,

White brodiaea

Death camas

Corn lily

shape, and often color. The three stamens are fused to the pistil, forming what may appear as a fourth petal, and the ovary is inferior. Orchid flowers attract pollinators by odor or nectar, and increase their cross-pollination success dramatically by remaining in bloom for long periods, fading only after fertilization has been achieved. Orchid seeds are tiny, with millions contained within each capsule. The seeds disperse with the wind like a fine powder, germinating wherever they can establish contact with the specific fungi required to supply necessary nutrients.

Phantom orchid (*Cephalanthera austiniae*)

• 8 to 20 inches
• mid-season
• semi-shady forest

This species is truly a phantom in the Basin, occurring only in years of plentiful moisture at low elevations along the western shore, from Homewood to Emerald Bay. Non-photosynthetic, its white flowers and stem appear ghost-like in the shady habitat in which it grows. This species is named for Rebecca Austin, an amateur collector who explored northeastern California from 1866 to 1900.

Spotted coral root (*Corallorhiza maculata*)

• 6 to 14 inches
• mid-season
• shady forest

This non-photosynthesizing orchid is occasional under the shady canopies of white fir and red fir forests. Coral root obtains carbohydrates and nutrients by tapping into fungal associations in the forest soils. It grows erect on a thin pink-reddish stem with small, similarly colored flowers whose lip petal is white with clear red-purple spots. The name derives from the coral-like shape of its root system. **Alaska rein orchid** (*Piperia unalascensis*) is uncommom in dry, shady to semi-open forest habitat.

Appearing as a skinnier version of bog orchid, it has two to three oblanceolate basal leaves and small green flowers, which hang on short, drooping pedicels from a narrow one- to two-foot, erect stem. **Rattlesnake plantain** (*Goodyera oblongifolia*) is occasional (sometimes locally abundant) under shady canopies of white fir forest along the western lake shore and around Echo Lakes. It has elliptic basal leaves with a white-striped midrib and many white-pink flowers blooming in a *raceme* on an erect stem.

Broad-leaved twayblade (*Listera convallarioides*)

• 4 to 12 inches
• mid-season
• moist, shaded streams, seeps

This delicate, rare orchid occurs under the tangled canopy along small, low-elevation streams. Confounding its scarcity, twayblade is easy to miss with its small, inconspicuous flowers in which the notched lip petal is a leaf-like green. The one- to three-inch-long, opposite, parallel-veined, ovate leaves are perhaps a better first indication of its presence. Another moisture-loving orchid now thought to be extinct in the Basin is **stream orchid** (*Epipactis gigantea*). This pleasing species historically grew at low elevations in the Basin and was last seen at Emerald Bay in 1944.

Phantom orchid

Spotted coral root

Rein orchid (*Platanthera leucostachys*)

- 8 to 30 inches
- mid-season
- wet to moist meadow, seeps, stream banks

Rein orchid is Tahoe's most common orchid, occurring from low elevations to the subalpine zone in a variety of moist habitats. The small, slightly fragrant white flowers form a spike inflorescence. Each flower has two spreading sepals with the third resting above in hood-like fashion. The two lateral petals are erect, while the lip petal extends downward with a distinct spur projecting out the back, nearly twice the length of the lip. The resemblance of the spur to a

Broad-leaved twayblade

horse's rein gives this species its common name.

Bog orchid (*Platanthera sparsiflora*)

- 8 to 30 inches
- mid-season
- wet to moist seeps, stream banks

Similar to rein orchid, bog orchid has more widely spaced, smaller green flowers in which the spur is approximately equal in size to the lip petal. Both orchids have linear leaves, which are reduced up the stem. Bog orchid is slightly less common than its white-flowered relative. The two species often occur together in moist habitats.

Ladies tresses (*Spiranthes romanzoffiana*)

- 4 to 10 inches
- late season
- wet meadows, stream banks

Ladies tresses is occasional in low- to medium-elevation locations such as Paige Meadows, Osgood Swamp, Tahoe Meadows, and Grass Lake. The linear leaves are reduced up the stem. It can be recognized by the whorls of densely packed, overlapping white flowers whose petals are fused together into slightly swollen tubes. Ladies tresses is often missed among the thick grasses where it usually blooms, late in the season. The rarer *Spiranthes porrifolia* has been found at Meeks Bay, Benwood Meadow, and Osgood Swamp. This species is twice the size of its more common relative, with slightly yel- lowish flowers whose lanceolate lip petals have tiny hairs at the tip.

DICOTS

The dicotyledons, or dicots, are characterized by seed embryos that have two initial leaves, flower parts typically in sets of four or five, and netted leaf venation. Dicots comprise the vast majority of flowering plants. Most of Tahoe's plant species are dicots.

MAPLE FAMILY (ACERACEAE)

The maple family is small, with only two genera and 120 species worldwide. It is closely related to the horsechestnut family, Hippocastanaceae, which includes California buckeye (*Aesculus californica*). The family consists of trees and shrubs that have opposite, palmately lobed leaves, small yellowish green flowers with eight stamens, and a two-chambered superior ovary. Many flowers lack female parts, a condition known as staminate. The fruit is a pair of winged achenes called a samara.

Mountain maple (*Acer glabrum* var. *torreyi*)

- 5 to 15 feet
- early season
- moist forest openings, slopes, stream edges

Rein orchid

Bog orchid

Tahoe's only maple representative, this species grows as a large shrub to small tree in moist habitats from low elevations up to 8,500 feet. The deciduous leaves have three main toothed lobes, the middle lobe being the largest. The stems are typically reddish. In fall the leaves turn various shades of pink to red, offering some complementary color to the golden cascades of aspens and cottonwoods. Big-leaf maple (*Acer macrophyllum*), common in the lower montane forest community, does not cross the western Sierra crest into the Basin.

Ladies tresses

CARROT FAMILY (APIACEAE)

Members of this large family, also called Umbelliferae, are known for their umbel flower arrangement in which three or more individual flowers radiate out on separate stalks from a central point. In Tahoe all genera (except *Sanicula*) possess a uniform, double umbellate structure in which the individual inflorescences together form their own umbels from a common node. The stems are hollow, and the leaves are alternate and often finely dissected. Each flower has five free petals, five stamens, one pistil, and an inferior ovary. While taxonomic distinctions within Apiaceae can be subtle, most Tahoe species can be recognized by easily observable characters. Common crop plants in this family are carrots, parsnips, celery, fennel, parsley, dill, coriander, cumin, caraway, and anise.

Brewer's angelica (*Angelica breweri*)

• **3 to 5 feet**
• **late season**
• **moist to dry, open slopes, forest**

This large plant is common up to the subalpine zone, sometimes forming almost exclusive flowering stands in the waning weeks of the growing season. It has a white,

open inflorescence, and twice-pinnate leaves with three- to four-inch lanceolate, serrated leaflets. Angelica was named for its supposed angelic healing powers. The larger **poison hemlock (*Conium maculatum*)** grows up to eight feet tall in disturbed areas. Native to Europe and highly toxic, an extract from this species was the poison used to execute Socrates. It can be identified by its fern-like leaves and its distinctly purple-spotted stems.

Cymopterus (*Cymopterus terebinthinus* var. *californica*)

• **6 to 24 inches**
• **early season**
• **open, rocky slopes, forests**

This lacy-leaved, yellow-flowered plant is common from mid-elevations through the subalpine zone. It was formerly classified under the genus *Pteryxia*, a name derived from the Greek word for fern, in reference to its intricately dissected leaf structure. **Sierra podistera (*Podistera nevadensis*)** is a rare high-elevation species that blooms early near the summit of Freel Peak and Jobs Sister. The cream to yellowish flowers rise in a tight cup-like inflorescence about two inches above the compact, cushion-forming herbage. The twice-pinnate basal leaves have linear to lanceolate, pointed leaflets, resembling those of cut-leaved daisy, another local high-elevation species.

Mountain maple

Brewer's angelica

Cymopterus

Cow parsnip

Cow parsnip (*Heracleum lanatum*)

- **3 to 7 feet**
- **mid-season**
- **moist, semi-shaded forest**

Cow parsnip is easily identifiable by its palmately veined, serrated leaves, which may reach up to two feet in width, and by its large (up to a foot wide) umbel inflorescence of small white flowers. The thick stem is typically branched above. It occurs throughout the Basin up to 8,000 feet elevation. The genus is named in honor of Hercules.

Gray's lovage (*Ligusticum grayii*)

• 8 to 30 inches
• mid-season
• moist to dry meadow edges, forest openings

Gray's lovage is abundant from lake level through the subalpine zone. It has small white flowers and lacy, twice-pinnate leaves with narrow, acute leaflet segments. Typically, all but one of its leaves are basal. Its roots were highly prized as a food source by Native Americans.

Fern-leaved lomatium (*Lomatium dissectum* var. *multifidum*)

• 1 to 3 feet
• early season
• dry, open slopes, forest

This species is recognizable by its small yellow to maroon-colored flowers and twice-pinnate leaves with oblong, rounded leaflet segments. It occurs in semi-dry forest openings in the north Basin, and is particularly common along the east Sagehen Creek trail. Native Americans used the root as a remedy for a variety of respiratory illnesses. **Sierra lomatium (*Lomatium nevadense*)** has white to cream-colored flowers and more crowded, dissected leaves with slightly pointed leaflet segments. This east-side migrant occurs in the far north and south Basin amidst open sagebrush scrub and on rocky volcanic slopes up to 9,000 feet.

Spindle orogenia (*Orogenia fusiformis*)

• 1 to 6 inches
• very early season
• moist, rocky meadow edges

Tahoe's earliest bloomer, this uncommon plant occurs on the edges of snowmelt-fed meadows at lower-elevation northern locations such as Goose Meadows or Sagehen Creek. It has small white flowers that rise only a few inches off the ground and dissected leaves with linear leaflet segments.

Western sweet cicely (*Osmorhiza occidentalis*)

• 1 to 3 feet
• mid-season
• semi-shady forest openings

This plant is common throughout the upper montane zone. The wide, twice-pinnate leaves have oblong to lanceolate, serrated leaflets. The yellow flowers are small and inconspicuous. The fruits are long and cylindrical. The genus name means scented root. It is also called sweet-anise for its licorice odor. The less conspicuous, six- to twelve-inch **sweet cicely (*Osmorhiza chilensis*)** has white flowers and grows in moist, shady forests below 7,500 feet. The smaller, delicate, twice-pinnate leaves have serrated, ovate leaflets. Sweet cicely is

Gray's lovage

Fern-leaved lomatium

Spindle orogenia

Western sweet cicely

also native to southern South America. The rare **western oxypolis** (*Oxypolis occidentalis*) occurs in wet, marshy habitat near Echo Summit and at Osgood Swamp. This two- to four-foot-tall plant has a white, clustered inflorescence and large, hairless, pinnate leaves with serrated leaflets ranging from narrowly lanceolate to widely ovate.

Sierra yampah (*Perideridia parishii* var. *latifolia*)

- 8 to 20 inches
- mid-season
- moist to dry habitats

Also known as Sierra Queen Anne's lace, this delicate species is one of the signature plants of rocky meadows and well draining seeps from low elevations into the sub-alpine zone. The tiny white flowers cluster in delicate inflorescences that are borne atop ascending, equal-sized umbel rays (*peduncles*). The erect stems are narrow and wispy. The few leaves are divided into one to three pairs of long, linear leaflets that project upward like ghoulish fingers. **Lemmon's yampah (***Perideridia lemmonii***)** is less common in low-elevation, west side meadow habitats, often with Sierra yampah. It can be distinguished by its more spreading, unequal-sized *peduncles*. **Bolander's yampah (***P. bolanderi***)** has a greater number of shorter (less than 6 cm.) leaflets. The roots are short and radish-like, while those of the other two species resemble small carrots. Yampah roots were sought by Native Americans for their sweet, nutty flavor. The seeds have a caraway aroma and can be used as seasoning. Yampah is the Shoshone word for this plant.

Tuberous sanicle (*Sanicula tuberosa*)

- 1 to 8 inches
- early season
- moist to dry habitats

Tahoe's sanicles are small, yellow-flowered perennials that do not bear their flowers exclusively in double umbels (single umbel inflorescences are best observed on mature plants.) This species blooms early with violets and buttercups in a variety of habitats up to 8,500 feet. The leaves are one- to two-pinnate with deeply lobed, narrow triangular leaflets that are slightly purple-tinged on the edges. The stems are also reddish purple. The minuscule flowers may be either bisexual or staminate. On bisexual flowers, the yellow styles extend outward well past the flower corolla. **Sierra sanicle (***Sanicula graveolens***)** is common on open, dry flats in the far north and along the Carson Range from Genoa Peak south to Carson Pass. Slightly larger, this species is best distinguished by its conspicuous, ovate leaves with lacy, lobed leaflets. The strong odor of the leaves gives the plant its species name.

Ranger buttons (*Sphenosciadium capitellatum*)

- 2 to 5 feet
- mid- to late season
- wet meadows, stream banks

This species is more common in

Sierra yampah

Tuberous sanicle

Ranger buttons

compacted, spherical inflorescences, which project upward from the umbel center like orbiting planets. (The species name may be loosely translated as "umbrella of small heads.") The leaves are similar to those of Brewer's angelica, although smaller and more reduced up the stem. Like poison hemlock, ranger buttons is toxic.

DOGBANE FAMILY (APOCYNACEAE)

This large, mostly tropical family is poorly represented in California, with only three native genera. It includes the ornamentals *Nerium* (oleander) and *Plumeria*, as well as the naturalized periwinkles, *Catharanthus roseus* and *Vinca major*. Closely related to the milkweeds, this family has milky sap, entire, generally opposite leaves, and radial flowers with five semi-fused sepals and petals, five stamens, and

the south Basin, occurring from low elevations to the lower subalpine zone. It is named for its tightly

two superior ovaries. The alkaloids in this family are employed today for a variety of heart-related illnesses. Many species are poisonous to livestock.

Spreading dogbane (*Apocynum androsaemifolium*)

• **6 to 16 inches**
• **mid-season**
• **dry slopes, open forest**

Spreading dogbane is common into the lower subalpine zone, often amidst montane chaparral and Jeffrey pines. The large, round leaves are dark green with distinct white veins on the upper surface. The white to pink bell-shaped flowers are perfectly shaped for honeybees, a preferred pollinator. The flowers bloom in a cyme inflorescence, either erect or hanging. The name comes from the historical belief that the plants were poisonous to dogs.

MILKWEED FAMILY (ASCLEPIADACEAE)

This widespread, primarily tropical family has over 300 genera and 2,800 species worldwide, but only five genera in California. Milkweeds are named for the milky white sap that exudes from the plants' stems. The scientific name comes from Asklepios, the Greek god of medicine. Monarch butterfly caterpillars obtain alkaloids from milkweed plants, which make them unpalatable to hungry predators. This chemical protection is retained by the adult butterflies following metamorphosis. Milkweeds have been used historically in a variety of ways, including food flavoring, chewing gum, medicinal glue, rope and thread fibers, pillow stuffing, and buoyancy filler in life jackets.

Purple milkweed (*Asclepias cordifolia*)

• **1 to 2 feet**
• **mid-season**
• **dry, open slopes**

This common foothill species is occasional in the northwest Basin below 7,500 feet elevation. The peculiar flower structure, characteristic of the family, consists of five stamens fused into a filament column, with five attached circular appendages called hoods, and an anther head that surrounds the enlarged stigma in the center of the flower. The five sepals are reflexed, and the five purple petals are open. There are two superior ovaries, one of which typically aborts. The fruit is a follicle containing many flat seeds with silky hairs that aid in wind dispersal. The large, opposite leaves are cordate and clasp the main two-foot stem. Two other common eastern Sierra species, narrow-leaf milkweed (*Asclepias fascicularis*) and showy milkweed

Spreading dogbane

(*A. speciosa*), have not been found in the Basin.

SUNFLOWER FAMILY (ASTERACEAE)

The sunflowers are the largest angiosperm family, with over 20,000 species in approximately 1,300 genera worldwide. As with orchids, the key to the sunflowers' success has been the development of a specialized reproductive strategy, one well suited to their radiation from tropical environments into the drier temperate regions where they have flourished.

Inspection of a typical sunflower reveals that the flower head is composed of many individual flowers, usually of two kinds. In the middle are the *disk* flowers: small, narrow, typically five-lobed tubes, each with four to five stamens and a single two-branched pistil. On the edges of the receptacle are

Purple milkweed

borne the *ray* flowers, with a petal-like corolla that extends well beyond the minuscule flower tube. Ray flowers are usually either *pistillate* (lacking stamens) or completely sterile.

Flower heads with ray and disk flowers are called *radiate*. A few genera, such as *Cirsium, Antennaria,* and some species of *Erigeron,* have only disk flowers, a condition known as *discoid*. In other genera, such as *Agoseris, Taraxacum, Crepis,* and *Microseris,* flower heads are composed solely of ray-like flowers, a condition known as *ligulate*. In contrast to true ray flowers, ligulate flowers are bisexual, with both stamens and a pistil. The top of the inferior ovary usually bears a set of scales, bristles, or feathery plumes known as a *pappus*. Each flower produces a single small seed, protected within a hardened fruit known as the *achene*, of which unshelled sunflower seeds are a good example. Enclosing the entire flower receptacle is a set of bracts called *phyllaries*, collectively referred to as the *involucre*.

The development of this unique floral structure probably offered members of the sunflower family many advantages as ancestral species migrated into increasingly arid areas from their tropical origins. In these temperate regions, species diversity tended to be lower but with a greater number of individuals of the same species in a given area, thus reducing the importance of specific pollinator relationships. Instead, scarce resources could be conserved by producing small flowers that collectively attract many different pollinators. Development of the protective achene surrounding each seed made sunflower plants particularly well suited to avoiding insect predation. Once mature, seeds spread across the landscape, often with the aid of various pappus structures, the airy plumes of the many wind-dispersed species occurring in the Basin being just one example. The successful result of these adaptations can be seen in summer-drought, Mediterranean-type re-

gions such as California, where sunflowers represent fifteen percent of the state's vascular plant species. In drier southern California, over twenty percent are in the sunflower family. Not including the grasses, the vast majority of non-native, naturalized angiosperm species in California are members of the sunflower family, mostly from the dry Mediterranean regions of Europe and the Middle East. In the Tahoe Basin, Asteraceae is the largest family, with over eighty species representing more than forty genera.

Yarrow (*Achillea millefolium*)

- **6 to 18 inches**
- **early to mid-season**
- **open habitats**

Yarrow is abundant in a variety of habitats up to the alpine zone. It has small white to pink ray flowers and white to yellow disk flowers. The alternate leaves are finely pinnate. Yarrow occurs throughout the northern hemisphere, with many different varieties. Studies comparing sea-level plants to those at higher elevations reveal significant differences in physiological adaptations. Such research has been valuable in piecing together the process by which isolated populations may evolve into distinct species over time. The generic name is derived from Achilles, who

Yarrow

Western eupatorium

cured the wounds of his warriors with yarrow leaves. Native Americans also used the plant to treat injuries. The species name means thousand-leaved.

Western eupatorium (*Ageratina occidentalis*)

• 1 to 2 feet
• late season
• moist to dry, rocky slopes

Western eupatorium is common at middle to high elevations, primarily in the southern and northern ends of the Basin. It has alternate, triangular, serrated leaves and heads of small white to pink disk flowers. Eupatorium is named after a Greek king who is said to have discovered its historical use as an antidote for some kinds of poison.

Orange-flowered agoseris (*Agoseris aurantiaca*)

• 4 to 20 inches
• mid-season
• semi-moist forest openings

This uncommon plant occurs from lake level to the beginning of the subalpine zone. Characteristic of *Agoseris,* the flower heads occur singly on upright, leafless, branchless stems. Each flower head contains numerous ligulate flowers and is bordered by phyllaries in two to four overlapping series. This species is easily recognizable by its striking flower color.

Pale agoseris (*Agoseris glauca* var. *monticola*)

- 2 to 16 inches
- mid-season
- moist to dry meadows, rocky slopes

Pale agoseris occurs in a variety of habitats up to 10,000 feet. This variety has pale green, variable shaped, basal leaves, conspicuous yellow flowers, and a red-marked involucre. More common in the south Basin, *Agoseris glauca* var. *laciniata*, has longer, tapered leaves with irregular lobes that angle slightly toward the leaf base. This species can be distinguished by the thin shaft (referred to as the *beak*) that connects the fruit body (achene) to the pappus above. On pale agoseris, the beak is shorter than the achene-fruit. Occasional in semi-open, dry forest or chaparral, **spear-leaved agoseris** (*A. retrorsa*) has long, narrow, dark green leaves with tapered lobes that angle sharply back towards the leaf base. The *involucre* is long and cylindrical. The achenes are distinctly flat-topped and considerably shorter than the beak. **Large-flowered agoseris** (*A. grandiflora*) occurs in moist to dry, grassy areas. It grows to almost three feet in height with pale bluish green herbage and ephemeral yellow flowers. The top of the achene tapers gradually to the long beak. The annual **woodland agoseris** (*A.*

heterophylla) blooms early in low-elevation, drying meadows, often in large groups of two- to four-inch-tall plants. This species has small, entire, oblanceolate leaves and sparse, dark purple hairs on its phyllaries.

Pearly everlasting (*Anaphalis margaritacea*)

- 8 to 40 inches
- mid- to late season
- semi-moist forest openings, rocky habitats

This species is occasional from low elevations into the subalpine zone, most commonly in drying stream beds. It is named for the lasting quality of its flower heads when dried. Each flower head has pearly white, papery phyllaries clustered around narrow, yellow disk flowers. The lanceolate leaves are alternate and not significantly reduced up the stem.

Rosy everlasting (*Antennaria rosea*)

- 4 to 16 inches
- early season
- semi-moist, rocky habitats

Antennaria is characterized by the *dioecious* condition (separate male and female plants), papery flower heads, and woolly leaves greatly reduced in size up the stem. Species often grow in mats from underground *stolons*. Rosy everlasting occurs into the alpine zone in a va-

Orange-flowered agoseris

Pale agoseris

Pearly everlasting

Rosy everlasting

riety of open habitats. It can be distinguished by its rose-tinted, acute phyllaries, and white disk flowers. In the western states this species consists of only female plants, which reproduce asexually by seed.

Alpine everlasting (*A. media*) is a high-elevation species that occurs on moist, rocky terrain and in subalpine meadows. It has dark brown to black phyllaries, densely woolly, spoon-shaped leaves, and grows to six-inches tall. In the far north and south Basin, the matted **dwarf everlasting** (*A. dimorpha*), blooms early off one-inch stems in drying meadows and flats.

Heartleaf arnica (Arnica cordifolia)

• **6 to 12 inches**
• **early to mid-season**
• **dry forest openings**

The arnicas are a large and complex genus with many recorded species in Tahoe. The leaves are opposite, and the large, yellow, radiate flower heads have well developed disk and ray flowers. The identity of particular species in Tahoe is complicated by the possibility of hybrid or asexual populations that elude definitive characterization. Common up to 8,000 feet, heartleaf arnica can be easily identified by its two to four pairs of shallowly toothed leaves, which are strongly cordate at the base. There are one to five flower heads, and the pappus is white.

Seep-spring arnica (*Arnica longifolia*)

• **10 to 30 inches**
• **mid-season**
• **moist, rocky slopes, stream margins**

This common species forms dense patches on moist, seepy, volcanic or metamorphic soils from 7,000 feet through the subalpine zone. The five to seven pairs of lanceolate, green leaves have entire to slightly toothed margins and short, stiff hairs that make the plant slightly rough to the touch. The five to ten flower heads have acute phyllaries and a red to yellow-brown pappus. The less common **meadow arnica** (*Arnica chamissonis* ssp. *foliosa*) occurs in wet-boggy to meadowy habitats, usually at lower elevations in the south Basin. The narrow, lanceolate leaves have long, spreading hairs that give the plant a grayish green appearance. The five to ten flower heads have obtuse phyllaries with an extra tuft of hair on their inner tips. The pappus is white to straw-colored. Both species generally reproduce asexually from spreading rhizomes.

Soft arnica (*Arnica mollis*)

• **8 to 24 inches**
• **mid-season**
• **moist meadows, stream margins, slopes**

Soft arnica is common in moist

Heartleaf arnica

Seep-spring arnica

Soft arnica

habitats into the subalpine zone. It has three to five pairs of soft, hairy, sometimes silver-green, untoothed

leaves and one to three large, hemispheric flower heads with a yellow to brown pappus. Native Americans traditionally used arnica flowers to treat sprains, bruises, and external wounds.

Sierra arnica (*Arnica nevadensis*)

- **6 to 12 inches**
- **mid-season**
- **semi-moist forest openings, rocky slopes**

Common up to 10,000 feet, this species has two to three pairs of rounded, untoothed leaves and one to three flower heads. It is known to hybridize with *Arnica cordifolia*. (An asexual form of *A. nevadensis* was previously referred to as *A. tomentella*.) More common in northern California, **mountain arnica** (*A. latifolia*) occurs in similar habitats in the Desolation Wilderness and from Ward Valley to Donner Summit. It has two to four pairs of lanceolate to subcordate, toothed leaves. In contrast to *A. nevadensis* and *A. cordifolia*, the leaves grow *sessile* (without *petioles*) on the upper part of the stem. There are usually one to three flower heads.

Nodding arnica (*Arnica parryi*)

- **6 to 24 inches**
- **mid-season**
- **moist meadows, stream margins, slopes**

Nodding arnica shares the general habitat and elevation range of soft arnica throughout the Basin. It may be distinguished by its green, sparsely hairy leaves and smaller ray flowers, less than fifteen millimeters long. The three to nine flower heads nod in bud prior to blooming.

Mountain sagebrush (*Artemisia tridentata* ssp. *vaseyana*)

- **1 to 3 feet**
- **mid- to late season**
- **dry slopes, forest openings**

This *polyploid* subspecies is common from low elevations into the subalpine zone, eventually transitioning into the state plant of Nevada, (*Artemisia tridentata* ssp. *tridentata*) as one moves east to-

wards the Great Basin. In finer volcanic soils, its widely dispersed root system monopolizes available moisture, resulting in extensive stands. Its relative abundance in these habitats may be aided in part by alleopathy, in which plants exude toxins that inhibit the germination of nearby seedlings. Mountain sagebrush can be recognized by its fragrant, gray-green leaves, which are hairy and distinctly three-lobed. The disk flowers are mostly hidden within small, wind-pollinated flower heads that hang off many tall, erect stems.

The eight- to twenty-inch-tall **timberline sagebrush** (*Artemisia rothrockii*) is less common, occurring above 8,000 feet on rocky slopes and along meadow edges, mostly in southern parts of the Basin. It can be recognized by its sticky, resinous stems and leaves. **Snowfield sagebrush** (*A. spiciformis*) is uncommon in rocky, high-elevation meadows and on slopes in the south Basin. It has three- to six-lobed leaves and a preference for moister habitats. **Low sagebrush** (*A. arbuscula*), occurs in matted form above 8,500 feet on wind-blown summits and ridges. Like all sagebrushes, this four- to twelve-inch-tall species is strongly aromatic because of the presence of terpenoids in the leaves and inflorescence that inhibit herbivory.

Sierra arnica

Nodding arnica

Mountain sagebrush

Mugwort (*Artemisia douglasiana*)

- 1 to 4 feet
- late season
- moist to dry meadows, slopes, stream beds

This late bloomer occurs in semi-moist habitats up to 8,000 feet. It has narrowly elliptic, three- to five-lobed, green leaves and small, bell-shaped flower heads that nod in clusters on erect stalks. **Boreal sagewort** (*Artemisia norvegica* var. *saxa-tilis*) is a higher-elevation one- to three-foot-tall species that occurs on wet, rocky soils above 8,000 feet. It blooms abundantly at Carson Pass and less frequently in the north. The green leaves are one- to two-pinnately divided with narrow leaflets. The nodding, yellowish flower heads have dark-margined phyllaries. **Silver wormwood** (*A. ludoviciana* ssp. *incompta*) has smaller flower heads and somewhat transparent phyllary margins. The deeply lobed leaves have white matted hairs, especially on the underside, and a sweet lemon odor, akin to the various species of *Ericameria*. It occurs on moist, rocky slopes near Mount Rose and at Carson Pass between 8,500 and 9,500 feet.

Alpine aster (*Aster alpigenus*)

- **4 to 16 inches**
- **mid-season**
- **wet to moist, rocky meadows**

Aster means star in Latin, a reference to the radiate flower heads of these plants, which typically contain many yellow disk flowers and white to purple ray flowers. Asters are notoriously difficult to distinguish from the closely related daisies in the genus *Erigeron*. In general, asters bloom later in the summer, from mid-July onward, with many flower heads and different-sized phyllaries, usually in indistinguishable rows. Daisies bloom earlier with fewer flower heads, and similar-sized phyllaries in more distinguishable rows. The common alpine aster occurs most frequently in subalpine meadows. Uncharacteristic of the genus, its large flower heads, with white to pink ray flowers, are always solitary

on a nearly leafless stem, the upper part of which is covered with white woolly hairs. The grass-like basal leaves are linear and lack stems.

Brewer's golden aster (*Aster breweri*)

- **1 to 3 feet**
- **mid- to late season**
- **semi-shady forest, open slopes**

Formerly *Chrysopis breweri,* this atypical species is common in semi-moist to dry habitats up to 10,000 feet. Highly variable, Brewer's golden aster may grow as a single-stemmed herb or as a semi-woody, multi-stemmed shrub, displaying many small yellow *discoid* flower heads, which lack ray flowers.

Wavy-leaved aster (*Aster integrifolius*)

- **8 to 30 inches**
- **mid- to late season**
- **dry forest openings, slopes**

This distinctive aster is common from low elevations into the subalpine zone. The upwardly reduced leaves are elliptic to obovate. The flower heads have deep violet to purple, slightly twisted ray flowers that are arranged in a haphazard manner around yellow disk flowers atop a semi-cylindrical involucre. It is best identified by its strongly glandular upper stems and phyllaries. An infrequent, dry

Mugwort

Brewer's golden aster

Wavy-leaved aster

Alpine aster

low elevations. It has reddish brown, hairy stems and elliptic to obovate leaves, which are distinctly serrated and somewhat rough to the touch.

habitat species is **round-leaved aster** (*Aster radulinus*), which grows to two feet in forested openings at

Western aster (*Aster occidentalis* var. *occidentalis*)

- 8 to 30 inches
- mid- to late season
- moist to dry meadows, forest openings

Western aster is common to slightly over 9,000 feet. The many small flower heads have violet-blue ray flowers and acute, pale-margined phyllaries. The linear to elliptic leaves are gradually reduced upward, and the stems are either hairless or have hairs that tend to grow in lines. **Ascending aster** (*Aster ascendens*) occurs in moist meadows and other grassy areas below 9,000 feet. It is shorter (8 to 24 in.), with stiff, short-hairy flower heads and more obtuse, greener phyllaries. Leaves are oblanceolate to oblong. The two- to three-foot-tall **Eaton's aster** (*A. eatonii*) (see photo on back cover) occurs amidst willows and alders in northeast wetland habitats. It has long, hairless, lanceolate leaves, which are not reduced up the stem (the mid-stem leaves are usually the largest), and large, pink ray flowers, which bloom mid- to late season. **Oregon aster** (*Aster oregonensis*) is occasional along low-elevation open forest trails around Shirley Canyon and Glen Alpine. Flower heads generally have five small white ray flowers and numerous disk flowers with distinctly purple anthers.

Arrow-leaved balsam-root (*Balsamorhiza sagittata*)

- 1 to 3 feet
- early season
- open forest, slopes

Common into the subalpine zone, this species has soft, finely hairy, strongly cordate basal leaves, which resemble arrowheads. The few stem leaves are lanceolate and significantly reduced up the stem. These leaf characteristics distinguish arrow-leaved balsam-root from the similar, slightly later blooming woolly mules ears. The northern Nevada species, *Balsamorhiza macrolepis* var. *platylepis* occurs on dry volcanic soils in the vicinity of Prosser Reservoir. It has silky, pinnate leaves and ovate phyllaries. Native Americans used the pounded roots of balsam-root as a salve on open wounds.

Large-flowered brickellia (*Brickellia grandiflora*)

- 12 to 30 inches
- mid- to late season
- moist to dry, rocky, open habitats

This species has inconspicuous, yellow-green discoid flower heads tangled together on numerous loosely spaced stems. It occurs mostly in the south Basin, often along trails. It has lanceolate to triangular opposite leaves, which are slightly cordate and serrated, similar in appearance to the leaves of

Western aster

Large-flowered brickellia

Arrow-leaved balsam-root

Alpine pincushion

giant hyssop. The slightly smaller
Greene's brickellia (*Brickellia
greenei*) has been found along the
Eagle Falls trail at Emerald Bay. It
may be distinguished by its sticky,
glandular herbage and ovate, alternate leaves, which crowd the solitary flower heads.

Alpine pincushion (*Chaenactis alpigena*)

- 1 to 4 inches
- mid-season
- rocky slopes, summits

This alpine species occurs on rocky
plateaus, slopes, and summits
above 8,500 feet. It is best recog-

nized by its densely white-woolly, non-glandular phyllaries. Species in this genus are called pincushion for their five-lobed disk flowers, each with a delicately branched stigma that protrudes upward in two recurving arcs.

Dusty maidens (*Chaenactis douglasii*)

• **2 to 20 inches**
• **early to mid-season**
• **dry, disturbed areas, forests, slopes, summits**

This common species has two intergrading varieties occuring in Tahoe. The taller, lower-elevation *Chaenactis douglasii* var. *douglasii* grows erect on a single, upwardly branched, semi-leafy stem. The matted *C. d.* var. *alpina* (or alpine dusty maidens) occurs at higher elevations, with flower heads blooming on typically leafless stems. The twice-pinnate leaves are grayish green with leaf lobes that are longest near the middle of the elliptic leaf blade and distinctly curled along the margins and the tip. The flower heads have sparsely hairy, glandular phyllaries. The northern species, **Sierra pincushion** (*C. nevadensis*), occurs on dry, high-elevation slopes in the north, extending along the western crest southward to the Desolation Wilderness, where it increasingly hybridizes with *C. d.* var. *alpina* in their shared habitats. This species

has ovate leaf blades with flat-tipped lobes that are largest near the leaf base. The phyllaries are hairy and sparsely glandular.

Rabbitbrush (*Chrysothamnus nauseous*)

• **1 to 3 feet**
• **late season**
• **dry, open forest, slopes, ridges**

This dry-adapted shrub is abundant in a variety of habitats and elevations. The yellow flower heads, each typically containing five disk flowers, are narrow and densely packed. The phyllaries are strongly keeled. This species is highly variable in Tahoe, with leaves ranging from thread-like to linear and herbage ranging from grayish white to bright green. **Parry's rabbitbrush** (*Chrysothamnus parryi*) is a similar, sometimes prostrate species distinguishable by its weakly keeled phyllaries and five to eighteen flowers per head. Two subspecies, from Carson Pass and the summit of Mount Rose, have been found in Tahoe. **Yellow rabbitbrush** (*C. viscidiflorus* ssp. *viscidiflorus*) occurs on the dry western slopes of the Carson Range. This species is distinguishable by its hairless stems and wider (1 to 10 mm.), greenish leaves, which are generally twisted and also hairless, except for small straight hairs along the margins. *Chrysothamnus* is closely related to

Dusty maidens

Anderson's thistle

Rabbitbrush

Ericameria and may one day be included in that genus.

Anderson's thistle (*Cirsium andersonii*)

- 18 to 40 inches
- mid-season
- dry, open forests, slopes

This common species has bright reddish purple discoid flower heads that stand erect on single or branched stems. Like most species of *Cirsium*, its leaf lobes and outer phyllaries are spine-tipped, an oft repeated adaptation that discourages grazing herbivores. **Elk thistle** (***C. scariosum***) (previously *C. drummondii*) occurs in a variety of volvanic or metamorphic habitats ranging from low-elevation moist meadows and forests to the rocky talus summit of Mount Tallac. The creamy white, discoid flower heads are surrounded by spiny-lobed, densely woolly leaves. This species may occur prostrate or upright. A third native, which occurs on high volcanic slopes in the north and south Carson Range, has narrow, extremely long-spiny, multi-lobed leaves, which are white-tomentose on both sides, and dull white flower heads. Various taxonomic treatments refer to this plant as *C. canovirens* (Carson Pass) or the closely related *C. subniveum* (Mount Rose). The non-native **bull thistle** (*C. vulgare*) occurs in disturbed areas at low elevations. It has upright, urn-shaped flower heads with bright purple flowers and tiny, spine-like bristles on the leaf upper surfaces.

Hawksbeard (*Crepis acuminata*)

- 8 to 30 inches
- early to mid-season
- dry, open forests, slopes

Crepis is characterized by ligulate flower heads and (in contrast to the other dandelion-like genera such as *Agoseris*), branched stems, deeply lobed stem leaves, and two distinct sets of inner and outer phyllaries. Hawksbeard is most abundant on dry, open hillsides into the subalpine zone. This grayish green species grows erect with acutely lobed upper stem leaves and numerous small flower heads. The one centimeter long, narrow involucre has five to eight inner phyllaries and five to ten ligulate flowers. The similar *C. pleurocarpa* occurs in dry, high-elevation sagebrush scrub in the southern Carson Range. It may be distinguished by its overall gray appearance, hairy phyllaries, and reduced upper stem leaves.

Western hawksbeard (*Crepis occidentalis*)

- 4 to 12 inches
- early to mid-season
- dry, open forests, rocky slopes, ridgetops

This shorter, more compact plant is occasional in the north Basin up to 9,000 feet, becoming rarer toward Carson Pass. It grows semi-prostrate, with large flower heads (up to 2 cm. long) containing more than eight inner phyllaries and ten ligulate flowers. This variable species ranges in color from green to gray. The similar *Crepis modocensis* occurs on dry, high-elevation volcanic slopes in the north and south Carson Range. It differs in having long, dark hairs that grow on its white-woolly phyllaries.

White-stemmed goldenbush (*Ericameria discoidea*)

- 4 to 16 inches
- mid-season
- rocky ridges, summits

Ericameria is characterized by a heavily glandular, yellow inflorescence that exudes a pungent, lemony odor. The phyllaries are in unequal ranks. This species is a strictly high-elevation subshrub that occurs above 8,500 feet in rocky habitats. It has discoid flower heads, small ovate leaves, and dis-

Hawksbeard

White-stemmed goldenbush

Western hawksbeard

inch **cliff goldenbush** (*E. cuneata var. cuneata*) blooms very late in the season on steep, granite cliffs and ledges around Shirley Canyon. The gland-dotted, obovate leaves are wedge-shaped at the top. There are from zero to three ray flowers. Goldenbush previously was placed in the genus *Haplopappus*.

Single-head goldenbush (*Ericameria suffruticosa*)

• **6 to 16 inches**
• **mid- to late season**
• **rocky slopes, ridges, summits**

This woody subshrub is common at middle to high elevations. The many flower heads generally grow on separate stems with dark green, oblanceolate leaves that are folded inward. Each flower head has nu-

tinctive, whitish-woolly stems. **Bloomer's goldenbush** (*E. bloomeri*) grows as a small shrub in similar habitats. It has linear, threadlike leaves and numerous flower heads with four to thirteen disk flowers and one to five inconspicuous ray flowers. The two- to sixmerous disk flowers and from one to six haphazardly arranged ray flowers. The Great Basin species, **spineless horsebrush** (*Tetradymia canescens*), appears on dry, high-elevation slopes and plateaus in the southern Carson Range. Similar in appearance to *Ericameria*, it can be distinguished by its bright yellow flower heads, each containing four equal-sized phyllaries and four disk flowers.

Shining daisy (*Erigeron barbellulatus*)

- **2 to 8 inches**
- **mid-season**
- **rocky slopes, ridgetops**

Tahoe's most common high-elevation daisy occurs on rocky summit approaches from 8,000 to 10,000 feet. Shining daisy is named for the sunlit reflection produced by the many small hairs on its stems and leaves. It has finely hairy phyllaries and mostly basal, narrowly oblanceolate leaves with whitish, hardened, and slightly expanded bases. **Sierra daisy** (*Erigeron algidus*) grows to six inches tall in slightly moister rocky habitats in the south Basin. It has long-hairy, faintly glandular herbage, elongated, spoon-shaped basal leaves, and white to pink-blue ray flowers that tend to reflex and coil after bloom-

ing. The Great Basin native, **Eaton's daisy** (*E. eatonii*) is rare in Tahoe in dry, grassy to rocky habitats up to 8,500 feet. The well developed leaves resemble those of a desert willow species: narrowly oblanceolate, grayish green, and three-veined. The numerous ray flowers are white with pinkish blue undersides.

Great Basin rayless daisy (*Erigeron aphanactis*)

- **3 to 8 inches**
- **mid-season**
- **dry ridges, summits**

This high-elevation species has bright yellow discoid flower heads that are slightly raised and flat-topped, with one set of same-sized phyllaries. It occurs along the Carson Range, from Slide Mountain south to Hawkins Peak. The plant is long-hairy and glandular, with linear to oblanceolate leaves, which are reduced up the stem. The east side native **ground daisy** (*Townsendia scapigera*) occurs in the Red Lake Peak area near Carson Pass. The large, flattened flower heads (white to pinkish ray flowers and yellow disk flowers) bloom flat on the ground, almost completely covering the narrow, spoon-shaped leaves, which crowd the stem base.

Single-head goldenbush

Shining daisy

Great Basin rayless daisy

Brewer's daisy

Brewer's daisy (*Erigeron breweri*)

• 6 to 20 inches
• mid- to late season
• dry, open forests, slopes

This straggly daisy is best recognized by its many small, hairy, oblanceolate leaves, which are evenly spaced and not significantly reduced in size up the stem. It has one to six flower heads with bright, pink-purple ray flowers. It occurs most commonly in the south Basin on dry, south-facing slopes up to 9,000 feet.

Cut-leaved daisy (*Erigeron compositus*)

- 2 to 6 inches
- early season
- open, rocky slopes, summits

This common species is recognized by its ternately dissected (divided into three parts) basal leaves. Usually, the flower heads are radiate, with white (occasionally blue to pink) ray flowers, but many populations are discoid. Another rayless species, **starved daisy** (*Erigeron miser*), occurs in the far north amidst the granite out-croppings of Donner Summit. This two- to ten-inch plant has yellow flower heads and small oblanceolate cauline leaves, which are evenly spaced up the stem. The phyllaries are glandular. This species is endemic to the Tahoe National Forest. The one- to three-foot-tall **rayless daisy** (*E. inornatus* var. *inornatus*) has been found on dry, typically volcanic slopes at low elevations in the northeast Basin near Incline. It differs from the superficially similar *Aster breweri* in its smaller (7 to 12 mm. wide) flower heads and oblong leaves with rounded tips.

Coulter's daisy (*Erigeron coulteri*)

- 8 to 20 inches
- mid-season
- moist, semi-shady stream banks, seeps, meadows

This relatively common species has one to four flat-topped flower heads with numerous narrow, white, sometimes coiled ray flowers. The flower heads typically nod in bud. The oblanceolate to elliptic leaves clasp the erect stem. The similar **ox-eye daisy** (*Leucanthemum vulgare*) has large white ray flowers and tightly packed yellow disk flowers. This European native differs from asters and daisies in the absence of a pappus and the relatively small, pinnately lobed to toothed stem leaves. It is common in moist meadows and disturbed areas. The rare **Linear-leaved daisy** (*Erigeron linearis*) occurs on dry, rocky slopes and summits between 8,000 and 9,000 feet This four- to eight-inch species has grass-like leaves and small, solitary flower heads, each with twenty-five to forty yellow ray flowers.

Wandering daisy (*Erigeron peregrinus*)

- 6 to 30 inches
- early to mid-season
- semi-moist forest openings, rocky slopes

Tahoe's most abundant *Erigeron*, this species occurs in habitats ranging from moist, semi-shady forests and meadows to high-elevation, open talus slopes. The long lower leaves are oblanceolate, and the reduced upper leaves are lanceolate. There are one to four flower heads

Cut-leaved daisy

Coulter's daisy

Wandering daisy

ters by its long, acuminate, dark-tipped phyllaries, which are often reflexed and strongly glandular.

Dwarf alpine daisy (*Erigeron pygmaeus*)

• **2 to 4 inches**
• **early to mid-season**
• **rocky alpine summits**

This diminutive species is prostrate with variable-sized flower heads that bloom above small basal leaves, which are linear and strongly glandular. The ray flowers are an eye-catching, deep violet-purple. Dwarf alpine daisy occurs only in the immediate vicinity of the summits of Mount Rose and Freel Peak, the two highest points in Tahoe.

with white to purple ray flowers. Wandering daisy is best distinguished from similar-looking as-

Woolly sunflower (*Eriophyllum lanatum* var. *integrifolium*)

• 4 to 30 inches
• early to mid-season
• dry, open forest, rocky slopes, summits

This species is widely distributed in the Basin, from low-elevation forests to the highest summits. Like showy penstemon, it becomes shorter with increasing elevation. The whitish green, woolly leaves are occasionally lobed. The single yellow flower heads have eight to thirteen ray flowers, each with three small teeth at the corolla tip. Woolly sunflower has eight recognized varieties in California, which occur from sea level to over 12,000 feet.

Marsh gnaphalium (*Gnaphalium palustre*)

• 1 to 6 inches
• mid-season
• vernal pools, wet to dry meadows

This moisture-loving annual is densely woolly with discoid flower heads on low-branching stems. It is especially common in low elevation, vernal pool habitats. Somewhat similar in appearance to the everlastings, the biennial **cudweed** (*Gnaphalium canescens* ssp. *thermale*) occurs in open, dry, often-disturbed habitats below 7,500 feet. It differs in having a multi-branched stem system with many

white to straw-colored discoid flower heads.

Bigelow sneezeweed (*Helenium bigelovii*)

• 1 to 4 feet
• mid-season
• wet to moist meadows, stream banks

This species is occasional in medium-elevation locations such as Tahoe Meadows and Glen Alpine. It is distinctive for its large size and alternate, clasping leaves, which are oblanceolate to linear. The distinctly spherical flower head receptacles host many long, yellow, often three-lobed ray flowers and numerous slightly raised, yellow to reddish purple disk flowers. The *Helenium* pappus differs from those of similar genera in its five to ten flat, membranous structures called *scales*.

California helianthella (*Helianthella californica* var. *nevadensis*)

• 12 to 30 inches
• mid-season
• semi-moist to dry, open forests

California helianthella is occasional below 8,000 feet in the northwest Basin.(It has also been collected at Cascade Lake and Sagehen Creek.) The long, narrow leaves are dark green and three-veined. The yellow disk flowers are raised above the rays, creating a

Dwarf alpine daisy

Woolly sunflower

Marsh gnaphalium

Bigelow sneezeweed

distinctly flat-topped flower head. A hand lens reveals that this species has a pappus of two awl-shaped scales that sit opposite each other on the apex of the achene. The non-native **common sunflower** (*Helianthus annuus*) occurs in disturbed areas around Tahoe. This large-flowered annual can be most easily recognized by its lanceolate to ovate leaves, which are serrated with a cordate base.

Golden aster (*Heterotheca villosa* var. *hispida*)

- **4 to 20 inches**
- **early to mid-season**
- **rocky ledges, crevices**

Also known as telegraph weed, this plant is occasional at low- to middle-elevation locations such as Shirley Canyon and Emerald Bay. A variable species, it occurs in Tahoe as a sprawling to erect perennial, with small leaves that are alternate and glandular-hairy. The flower heads have yellow ray and disk flowers. The phyllaries are in three to five graded series. The one-to two-foot western plains native, **gumplant (*Grindelia squarrosa* var. *serrulata*)** grows as a weed throughout Tahoe, blooming late in the season along roadways or in other disturbed areas. Like many other *Grindelia* species, the radiate flower heads are brilliant yellow with glandular-sticky phyllaries in five to six series. In this species the phyllaries are long, tapered, and strongly recurved, forming a full backward circle with the phyllary base. The alternate, hairless leaves are dentate, with swollen tips, and oblong to ovate.

Shaggy hawkweed (*Hieracium horridum*)

- **4 to 12 inches**
- **mid-season**
- **open, sandy, rocky slopes**

Hieracium species are generally perennials with milky sap, small, ligulate flowers, and densely hairy lower stems and leaves. They differ from similar Tahoe species in their several to many flower heads per stem and their fruits, which lack any beak attachment to the pappus (compare with various species in *Agoseris*). Shaggy hawkweed is occasional in (typically) granitic terrain up to 10,000 feet. The leaves and stems are white-woolly. The delicate **alpine hawkweed (*Hieracium gracile*)** occurs on semi-moist, subalpine slopes and forest openings on the west side. The six-to ten-inch, mostly leafless stems bear several small yellow flower heads. The phyllaries have long black hairs. **White-flowered hawkweed (*H. albiflorum*)** is abundant in sparsely lit forest understories. The delicate white flower heads are borne on narrow one- to two-foot stems that rise erect above oblanceolate basal leaves. The similar **trail marker (*Adenocaulon bicolor*)** is occasional in low-elevation shady forest habitat along the west lake shore. This common foothill species has tiny white, *discoid* flower heads and large, triangular basal leaves.

California helianthella

Golden aster

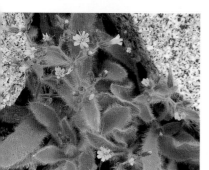

Shaggy hawkweed

Alpine gold

Alpine gold (*Hulsea algida*)

- 4 to 12 inches
- mid-season
- alpine summits, ridges

Alpine gold occurs on the high-elevation gravel scree and talus slopes of Mount Rose, the Freel Peak-Jobs Sister complex, and Round Top. This long-hairy, glandular plant has coarsely toothed basal leaves and large flower heads with yellow ray and disk flowers. *Hulsea heterochroma* has been found amidst mid-elevation montane chaparral below Mount Tallac. This tall (up to three feet), glandular species has coarsely serrated leaves and striking red ray flowers.

Hoary aster (*Machaeranthera canescens* var. *canescens*)

- 8 to 36 inches
- mid- to late season
- dry, sandy, rocky habitats

Hoary aster is occasional at low elevations, occurring frequently along the east-side lake shore. The multi-branched stems bear many small flower heads with pink to purple ray flowers and yellow disk flowers. It can be distinguished from *Aster* and *Erigeron* species by its highly glandular phyllaries with spreading tips, narrowly cylindrical involucre, and distinctly toothed stem leaves. A low-growing variety, *Machaeranthera canescens* var. *shastensis,* with large, magenta-purple flower heads, occurs on open, rocky slopes in the Carson Pass area and near Tinker's Knob.

Bolander's tarweed (*Madia bolanderi*)

- 2 to 5 feet
- mid- to late season
- wet meadows, seeps

This large perennial is occasional at low elevations. Like all tarweeds, it is densely glandular and aromatic with yellow flower heads whose ray flowers are strongly lobed. The flower heads have eight to twelve showy ray flowers and, unique among Tahoe tarweeds, pappus scales on each disk flower. The one- to three-foot annual, **elegant tarweed** (*Madia elegans* spp. *elegans*), occurs in semi-moist, well drained forest openings and rocky meadows. The attractive flower heads have five to twenty large ray flowers, a hairy receptacle, and infertile disk flowers.

Least tarweed (*Madia minima*)

- **1 to 5 inches**
- **mid-season**
- **forest openings, sandy flats, slopes**

This inconspicuous annual has thread-like, multi-branched stems, which support tiny flower heads with minute black glands, three to five ray flowers, and one to two fertile disk flowers. Tahoe is home to three other annual species that lack showy flower heads. **Mountain tarweed** (*Madia glomerata*) is common in the north in open, dry habitats in the upper montane zone. It grows to twenty inches. Each flower head has three short ray flowers and one to five fertile disk flowers, which form individual fruits that are widest in the middle. The similar sized **slender tarweed** (*M. gracilis*) differs in its three to nine larger ray flowers and fruits that are wide at the top and

Hoary aster

Least tarweed

gradually tapered to the base. **Threadstem madia** (*M. exigua*) has been found near Glen Alpine and at Donner Lake in dry habitats below 7,500 feet. This thin-stemmed species can be distinguished from least tarweed by its larger size (up to one foot) and by the prominent yellow glands that cover its flower heads.

Bolander's tarweed

Nodding microseris (*Microseris nutans*)

- **4 to 12 inches**
- **early to mid-season**
- **forest openings**

This common understory species has yellow, ligulate, solitary flower heads, which nod distinctly in the bud off narrow, erect to ascending stems. The leaves vary from linear to oblanceolate and entire to pinnately lobed. Nodding microseris differs from similar genera by the presence of lower stem leaves, sparsely hairy phyllaries, and pappus bristles whose widened bases form a tiny, umbrella-like saucer at the top of the fruit.

Alpine dandelion (*Nothocalais alpestris*)

- 2 to 10 inches
- early to mid-season
- semi-moist open forest, slopes, meadow edges

Alpine dandelion occurs from 7,000 to 9,500 feet, primarily in the southwest Basin. The ligulate flower heads bloom individually atop leafless stems. The highly variable basal leaves are linear to oblanceolate. The pappus is a single set of white, hair-like bristles. It differs from *Agoseris* species in its hairless stems, more or less equal-sized phyllaries (which are finely purple-speckled), and beakless achene-fruits, which attach directly to the pappus.

Woolly marbles (*Psilocarphus brevissimus* var. *brevissimus*)

- 1 to 2 inches
- mid-season
- drying vernal pools

This prostrate annual associates with vernal pool species such as porterella and needle navarretia in drying, low-elevation meadow habitats such as Pole Creek and Paige Meadows. It is densely woolly with short, spreading stems and marble-shaped, discoid flower heads whose hairy, leaf-like phyllaries conceal two to ten disk flowers.

Silver-leaf raillardella (*Raillardella argentea*)

- 2 to 5 inches
- mid-season
- dry, sandy to gravelly flats, slopes

This species is occasional in open subalpine to alpine habitats. The yellow, discoid flower heads are cylindrical to bell-shaped, each perching on its own erect to sprawling, leafless stem. The oblanceolate basal leaves are silver-green, hairy, and often grow in a rosette pattern. **Green-leaved raillardella (*Raillardella scaposa*)** has a more limited range, occurring above 9,000 feet on dry, open slopes in the vicinity of Mount Rose and at slightly lower elevations near Carson Pass. It differs in its sparsely hairy, glandular, green leaves.

Woolly groundsel (*Senecio canus*)

- 4 to 10 inches
- mid-season
- dry ridges, summits

Members of this genus have alternate leaves, one equal series of main phyllaries, and (typically) yellow flower heads, often with a reduced number of small ray flowers. One of several high-elevation species occurring in Tahoe, this gray-hairy, east-side species is occasional above 8,500 feet in the north and south Basin and especially common along the northern

Nodding microseris

Alpine dandelion

Woolly marbles

Silver-leaf raillardella

Carson Range. The lanceolate to ovate basal leaves are entire to slightly *dentate* (leaf margins with coarse teeth pointing outward).

Fremont's groundsel (*Senecio fremontii* var. *occidentalis*)

- **4 to 10 inches**
- **mid-season**
- **semi-moist, rocky slopes, rock crevices**

Fremont's groundsel is occasional on cool, semi-moist subalpine slopes, typically growing off a creeping rootstock from beneath large boulders, mostly in the south Basin. The dark green, dentate leaves are oblanceolate to ovate and evenly spaced (though reduced in size) up the curving, prostrate to ascending stem. There are one to five flower heads, each with eight ray flowers. The delicate, uncommon **Rocky Mountain groundsel** (*Senecio cymbalarioides*) occurs in damp, subalpine meadow habitats, growing erect up to a foot tall from a slender rhizome. The mostly basal leaves are somewhat fleshy and ovate, with entire to crenate margins. The single (occasionally up to three) flower head has thirteen ray flowers and blooms from mid- to late season.

Single-stemmed groundsel (*Senecio integerrimus*)

- **8 to 28 inches**
- **early season**
- **moist to dry forest openings, slopes**

This abundant species has a single upwardly branched stem and six to twenty flower heads usually containing around thirteen ray flowers. Two varieties occur in the Basin, distinguishable by the phyllary tips, which may be either green or strongly black-tipped. **Sierra groundsel** (*Senecio scorzenella*) is less widespread, growing on moist, grassy flats, meadow edges, and slopes up to 9,000 feet, usually in the southern Basin. The large, bluish green basal leaves are slightly fleshy and sparsely woolly on the underside with fine, sharp teeth along the margins. The flower heads are small with five or more thin ray flowers (5 to 8 mm. long) and dark-tipped phyllaries.

Arrowleaf groundsel (*Senecio triangularis*)

- **2 to 4 feet**
- **mid-season**
- **wet to moist, stream margins, meadows, seeps**

Arrowleaf groundsel is abundant throughout the Basin up to 9,000 feet. It has triangular, sometimes cordate, serrated leaves, which are only slightly reduced up the stem, and many small, individually stemmed flower heads, each with eight small ray flowers. Common in the Cascades, **meadow groundsel** (*Senecio serra* var. *serra*) can be locally abundant in low-elevation wet meadows in the north Basin. It grows to over six feet with smaller, well spaced, lanceolate leaves, which are reduced up the thick, whitish green, main stalk. The small flower heads cluster together in groups of five or more on single stems.

Woolly groundsel

Single-stemmed groundsel

Fremont's groundsel

Narrowleaf groundsel (*Senecio streptanthifolius*)

- 4 to 20 inches
- early to mid-season
- forest openings, rocky talus slopes, ridges

This variable Cascadian plant ranges from an erect forest dweller

Arrowleaf groundsel

to a prostrate alpine species over a wide variety of habitats in the far northern and southern Basin. The thick, mostly hairless leaves are bright green and, at high elevations, typically curled inward. The lower leaves are rounded to elliptic, dentate, and crowd the base of the stem. The occasional upper stem leaves are narrowly oblong and often deeply lobed. There are four to twenty flower heads. **Alpine groundsel** (*Senecio werneriifolius*) grows semi-prostrate, generally under four inches tall, on the high-elevation scree slopes of Freel Peak. The thick, long-stemmed, dark green basal leaves are slightly rolled under and have toothed margins.

Canada goldenrod (*Solidago canadensis* ssp. *elongata*)

• 1 to 4 feet
• mid- to late season
• moist to dry meadow edges, disturbed areas

This widespread plant occurs up to 8,000 feet in moist, grassy habitats. Characteristic of the genus, it has alternate, generally three-veined leaves, small flower heads with yellow disk and ray flowers and, in contrast to *Senecio,* phyllaries in three to five series. Canada goldenrod can be recognized by its toothed, lanceolate leaves, which are largest at mid-stem, and its tightly packed, club-shaped clusters of flower heads. The similar-sized **California goldenrod** (*Solidago californica*) is slightly less

common in similar habitats below 7,500 feet elevation. The herbage is densely short-hairy. The oblanceolate leaves are largest near the base of the stem and greatly reduced in size up the stem. The flower heads bloom in a one-sided, wand-shaped inflorescence.

Northern goldenrod (*Solidago multiradiata*)

• 6 to 20 inches
• mid- to late season
• semi-moist to dry, rocky ridges, slopes, meadows

This high-elevation species occurs in moist, rocky areas above 8,500 feet. Besides its distinct habitat preference, it can be distinguished by its relatively hairless lower herbage, smaller size, longer (4 to 7 mm.) involucre, and long, lanceolate phyllaries, which are not strongly graded.

Short-stemmed stenotus (*Stenotus acaulis*)

• 2 to 6 inches
• early to mid-season
• dry ridges, summits

This woody alpine shrub grows in mat form above 8,000 feet throughout Tahoe, particularly along the volcanic ridgetops characteristic of the northern and southern Basin. Its many yellow, radiate, solitary flower heads bloom on short stems above small, oblanceolate leaves.

Narrowleaf groundsel

Canada goldenrod

Northern goldenrod

Short-stemmed stenotus

Stephanomeria (*Stephanomeria lactucina*)

- 4 to 12 inches
- mid- to late season
- sandy forested openings

This attractive forest dweller is easy to miss because of its later blooming time and short period of flowering. There are seven to ten pink-lavender ligulate flowers atop erect, upwardly branched stems. The leaves are linear to lanceolate with occasionally toothed margins. **Narrow-leaved stephanomeria** (*Stephanomeria tenuifolia*) creeps into the Basin from the east on dry mountain slopes up to 10,000 feet. Also known as wire lettuce, it has slender, many-branched stems, almost thread-like leaves, and flower heads with five narrow, pink, ligulate flowers. **Thorny skeleton plant** (*S. spinosa*) occurs on dry volcanic slopes of the Carson Range, from Mount Rose south to Carson Pass. It has small, late-blooming, pink flowers and densely branched stems that dry to sharp spines, reminiscent of a desert species. Native Americans chewed the inside of *Stephanomeria* stems like gum.

Common dandelion (*Taraxacum officinale*)

- 4 to 16 inches
- early season
- disturbed areas, moist meadows

This abundant weed dots the early season landscape with flashes of bright yellow. It has a flat inflorescence containing numerous ligulate flowers and toothed to lobed leaves. It is best identified by its outer phyllaries, which are strongly reflexed back toward the stem.

Sierra tonestus (*Tonestus eximius*)

- 2 to 6 inches
- mid-season
- rocky slopes, summits, ridges

Sierra tonestus is occasional in rocky, open habitats above 8,500 feet. It grows as a three- to six-inch, non-woody perennial with one to several yellow flower heads and toothed to lobed alternate leaves, which are strongly glandular and hairy. Although widespread and not uncommon in Tahoe, Sierra tonestus is rare in California.

Salsify (*Tragopogon dubius*)

- 1 to 3 feet
- mid-season
- semi-moist to dry disturbed habitats

This striking European species is common at low elevations. It is dis-

Stephanomeria

Common dandelion

Sierra tonestus

Salsify

tinctive for its pale yellow, ligulate flower heads whose long acute phyllaries project considerably past the outermost ray flowers. Each inflorescence is solitary atop a tall stem that widens at the tip. The flower heads open in mid-morning and close by mid-day.

Woolly mules ears (*Wyethia mollis*)

• 1 to 2 feet
• early to mid-season
• open forest, slopes

The vast, open volcanic slopes covered by mules ears along the Pacific Crest Trail in the northern part of the Basin are a testament to this plant's local abundance in shallow, higher-elevation soils with some water-holding capacity. Common up to 8,500 feet, woolly mules ears has hemispheric flower heads and large, oblanceolate to obovate leaves that are covered with dense, matted hairs, giving the plant an overall grayish green appearance and silky texture. (*Mollis* means soft in Latin.) The leaves are distinguishable from those of the similar arrow-leaved balsam-root by their non-cordate bases, which instead taper gradually to the leaf mid-vein.

BIRCH FAMILY (BETULACEAE)

This mostly northern, temperate family consists of wind-pollinated, deciduous trees and shrubs with simple, alternate leaves and unisexual flowers that are borne on a hanging catkin inflorescence. The fruit consists of a small nutlet with winged bracts, which aid in wind dispersal. There are three genera in California, hazelnut (*Corylus*), birch (*Betula*), and alder (*Alnus*). Of the state's four alder species, only mountain alder occurs in the Basin.

Mountain alder (*Alnus incana* ssp. *tenuifolia*)

• 5 to 20 feet
• early season
• moist stream, lake, meadow margins

Mountain alder is abundant along stream courses up to 8,500 feet, often forming dense thickets on seepy north- or east-facing slopes. The ovate leaves have flat (not rolled) margins and a dark green upper surface with an indented midrib. Male and female flowers bloom on the same tree in late spring before the advent of leafy growth, a common adaptation that enhances wind pollination. The small woody cones persist on the stem into the next year. Mountain alder has root nodules containing nitrogen-fixing bacteria, which allow this species to flourish in the nitrogen-poor soils typical of its usual habitat. More common on the eastern Sierra, **water birch** (***Betula occidentalis***) has been found in the Carson Pass area. It grows as a small tree, with smaller, rounded, serrate, glandular leaves and non-woody catkin-like fruits.

Mountain alder

Woolly mules ears

Cryptantha

BORAGE FAMILY (BORAGINACEAE)

This family of mostly annual or perennial herbs has alternate, usually entire leaves and sharp, bristly hairs on stems and inflorescence. The flowers have five stamens and five partially fused petals that form a tube with spreading corolla lobes. Borage flowers can generally be recognized by the five raised appendages at the top of the flower tube, which help to preserve the nectar at the base of the ovary for long-tongued pollinators. The ovary is superior and usually four-lobed. The fruit is a nutlet.

Cryptantha (*Cryptantha affinis*)

- 4 to 12 inches
- early to mid-season
- sandy to rocky forest and chaparral openings

The most common of Tahoe's annual cryptanthas, this species has densely hairy herbage and small white flowers borne atop a single, sometimes upwardly branched stem. (The genus name is Greek for hidden flower.) The petals are rounded, and the leaves are oblong to lanceolate. *Cryptantha simulans* is similar with a branched stem and linear to narrowly oblanceolate leaves. Cryptanthas differ from members of *Plagiobothrys* in their drier habitat preference and their nutlet fruits, which have a recessed scar and a grooved nutlet wall. *Plagiobothrys* nutlets have an elevated scar and a keeled nutlet wall. Since these characteristics are best seen with a microscope, most hikers refer to the annual species of both genera as popcorn flowers.

Alpine cryptantha (*Cryptantha humilis*)

• **4 to 8 inches**
• **early to mid-season**
• **dry, rocky summits, ridges**

This perennial is occasional in dry, windswept habitats above 8,000 feet. It has oblanceolate to spoon-shaped, hairy, mostly basal leaves and a dense, cylindrical, many-flowered inflorescence. The flowers are white with yellow appendages at the top of the corolla tube. The similar Sierra cryptantha (*Cryptantha nubigena*) does not reach as far north as Tahoe.

Smooth stickseed (*Hackelia nervosa*)

• **16 to 32 inches**
• **early to mid-season**
• **dry, open slopes, forests**

Stickseeds are named for their prickly mature fruits, which adhere to any object, including the legs of passing hikers. (They are often called forget-me-nots due to their similarity to the related European genus *Myosotis*.) Smooth stickseed has sparse lower stem hairs, medium-sized flowers with corolla tubes that extend well past the calyx, and fruits with prickles covering the entire surface. **Velvety stickseed (*Hackelia velutina*) has larger (1 to 2 cm. wide) flowers and dense lower stem hairs. The similar northern species *H. californica*, rare in the north Basin, has white to slightly pinkish flowers. Jessica's stickseed (*H. micrantha*) has small flowers in which the corolla tube barely exceeds the calyx. The fruits have prickles in distinct lines, not covering the entire surface. The similar *H. floribunda*, found at Sagehen Creek and Carson Pass, has fruits with few to no facial prickles and marginal prickles that fuse at the base to form a wing. **Western hound's tongue (*Cynoglossum occidentale*) is rare in low-elevation forest openings in the north Basin. It has funnel-shaped, dull rose-reddish flowers, fruits without prickles and large, oblan-

Alpine cryptantha

Smooth stickseed

Mountain bluebells (*Mertensia ciliata*)

• 2 to 4 feet
• mid-season
• moist seeps, stream edges

Mountain bluebells

ceolate, sessile leaves, which are densely hairy and rough to the touch.

Mountain bluebells is occasional up to 9,000 feet in the south and east Basin. It has hairless, sometimes bluish green herbage with sessile to clasping, ovate to lanceolate leaves. The inflorescence hangs down in a *panicle* from a long ascending stem. Each flower has a small fused calyx and a long blue to pink-purple corolla tube with an exserted style and stigma. Members of this genus are also known as lungworts, due to the use of related species as a treatment for lung disease.

Plagiobothrys (*Plagiobothrys cognatus*)

- **1 to 3 inches**
- **early season**
- **moist meadows, vernal pools**

This complex genus consists of small, white-flowered annuals that are distinguished, as a genus and among individual species, by minute characteristics of the nutlets. The highly variable *Plagiobothrys cognatus* includes a common form in Tahoe that blooms in wet to moist grassy areas, often forming dotted white carpets across the spongy terrain. The tiny flowers sit atop a short, erect stem that has several pairs of small, elliptic leaves. *P. hispidulus* occurs in low-elevation meadows and along muddy lake margins. It has oblanceolate stem leaves and a coiled, many-flowered inflorescence borne along branching, prostrate to ascending stems. *P. hispidus* occurs in the east Basin in dry, sandy habitats up to 8,500 feet. More common in the sagebrush scrub to the east, this erect, two- to ten-inch annual has exceedingly bristly herbage and flowers.

MUSTARD FAMILY (BRASSICACEAE)

This northern temperate family has eleven genera and many species in Tahoe. The four-petaled flowers form a cross in the middle. (Another family name, Cruciferae, comes from the Latin for cross bearing.) There are six stamens, two shorter than the other four. The flowers are distinguished from those in the evening primrose family by their superior ovary, which develops above the flower petals into a variety of easily recognizable fruit shapes. The plants contain pungent oils, which give their leaves a spicy hot flavor. Common agricultural plants in this family include cabbage, turnip, cauliflower, rutabaga, brussel sprouts, broccoli, and radish.

Drummond's rock cress (*Arabis drummondii*)

- **1 to 3 feet**
- **early to mid-season**
- **dry to semi-moist open forest, rocky slopes**

Arabis is a taxonomically diverse, early blooming genus with many difficult-to-distinguish species in Tahoe. Rock cresses are typically dry-adapted, with small white to purplish flowers, and long, flattened fruits. Drummond's rock cress is occasional below 9,000 feet in the North Basin, often in slightly disturbed habitats. It has mostly hairless, slightly glaucous leaves and white to cream flowers that crowd the main stem tip. The fruits stand upright against the erect stem and generally carry the seeds in two

Plagiobothrys

Drummond's rock cress

Holboell's rock cress

vertical columns. The hybrid *A. divaricarpa* occurs in the south Basin on open, semi-moist, mid- to high-elevation slopes. This species has pink to purplish flowers, sparsely hairy, sometimes *dentate* leaves, and ascending to spreading fruits that carry the seeds in one

vertical column. The widespread **tower mustard** (*A. glabra* var. *glabra*) (also placed at times in the genus *Turritis*) is occasional in moist, grassy, often disturbed habitats, usually below 7,000 feet. This two- to four-foot species has large, oblanceolate basal leaves, cream to pale yellow flowers, and narrow, cylindrical fruits that grow erect against the stem.

Holboell's rock cress (*Arabis holboellii*)

• 1 to 3 feet
• early season
• dry, open forest, slopes

This abundant, low- to mid-elevation species has small white flowers, downward-hanging fruits, each with a single row of seeds, and lower leaves with small, multi-branched hairs. Tahoe's two varieties differ in their fruits and fruit stems, which are either straight or curved. This species is often attacked by a rust fungi, which appears as small orange bumps on the upper plant parts, preventing normal growth and flowering. The uncommon **bristle-leafed rock cress** (*Arabis rectissima* var. *rectissima*) also has hanging fruits and white to pinkish flowers, but differs in the erect, straight and forked hairs along its basal leaf margins. Also uncommon, **wavy-leaved rock cress** (*A. repanda* var. *repanda*) has white flowers, wide, ascending fruits, and large, oblanceolate basal leaves with wavy to *dentate* margins. The Great Basin species, *A. sparsiflora* var. *sparsiflora* occurs in the East Basin. It has large fruits that arch outwards from the stem like miniature sickles. The erect-stemmed *A. puberula* is occasional in the subalpine zone along the Carson Range. It has distinctive gray-green herbage, bright pink-purple flowers, straight hanging fruits, and many stem leaves with margins that are often wavy or toothed.

Lyall's rock cress (*Arabis lyallii*)

• **2 to 6 inches**
• **early to mid-season**
• **cliff sides, crevices, rocky slopes**

This attractive rock cress has typically hairless leaves and stems, spoon-shaped, rose to purple flowers, and straight, narrow, ascending fruits. It is occasional above 7,000 feet in rocky habitats from Donner Summit to Carson Pass. The similar **broad-seeded rock cress** (*Arabis platysperma*) is abundant from low-elevation forests to open slopes in the alpine zone. It is typically multi-branched, with small white to pink flowers and spoon-shaped petals. The distinctive fruits are three to five millimeters wide at maturity, flat and broadly spaced along several ascending stems. Tahoe's two varieties are distinguishable by the presence or absence of minute, multi-forked hairs on the basal leaves. **Lemmon's rock cress** (*A. lemmonii*) grows to ten inches on high-elevation, rocky slopes. It has purplish flowers, long, delicate, ascending stems, and typically narrow, spreading to hanging fruits. The basal leaves are covered with minute, dense, stellate hairs. The rare *A. tiehmii* (named after local botanist Arnold Tiehm) has been found in rocky habitats near Mount Rose. It has similar slender, ascending stems, which bear small white flowers and straight, narrow, upright fruits. The erect, generally hairless basal leaves have a distinct, straight hair at the tip.

Lyall's rock cress

Heart-leaf bitter cress

compound leaves, and long, cylindrical, erect to spreading fruits, which carry the seeds in a single column. The leaves may be used as spicy additions to a salad.

Heart-leaf bitter cress (*Cardamine cordifolia* var. *lyallii*)

• 8 to 18 inches
• early to mid-season
• moist stream, lake margins

Cardamine can be identified by its small white flowers and long, flattened fruits, which carry the seeds in a single column. This species is recognizable by its cordate leaves. It decorates narrow, shaded stream banks with bright white flowers in the north Basin. **Brewer's bittercress** (*C. breweri* var. *breweri*) oc-

Winter cress

Winter cress (*Barbarea orthoceras*)

• 1 to 2 feet
• early season
• moist meadows, stream banks

The relatively common winter cress has small yellow flowers with greenish yellow sepals, pinnately

curs throughout the upper montane in similar habitats. It differs in its pinnately lobed basal leaves and less showy flower petals.

Tansy mustard (*Descurainia incisa* ssp. *incisa*)

• 1 to 3 feet
• early to mid-season
• moist to dry forests, meadows, disturbed habitats

This common, low-elevation annual has pinnately lobed leaves and tiny yellow flowers that hover delicately erect on a single, upwardly branched stem. The slightly curved, cylindrical fruits project upwards from narrow, spreading to ascending stems. The biennial *Descurainia incana* occurs in a variety of habitats, from moist meadows to dry volcanic summits. It has straighter fruits, upward-pointing fruit stems, and, at higher elevations, glandular and grayish herbage. **California tansy mustard** (*D. californica*) has been found in similar habitats in the east Basin up to 9,300 feet. It has smaller, elliptic to ovate fruits, which are widest in the middle and tapered at both ends.

Tahoe draba (*Draba asterophora* var. *asterophora*)

• 2 to 8 inches
• early season
• dry, rocky slopes, summits

Drabas are typically found at high elevations, growing in mats or cushions that protect against constant windy conditions. Tahoe draba has dark green, round to obovate leaves that are five to fifteen millimeters long and covered with small, four-branched (*stellate*) hairs. This variety occurs on north or east facing slopes in the Mount Rose area and in the south from Monument to Freel Peak. **Cup Lake draba** (*Draba asterophora* var. *macrocarpa*) has longer styles (1 to 2 mm.) and occurs above Cup Lake and along ridges east of Ralston Peak. Common to the south, **Lemmon's draba** (*D. lemmonii*) occurs on the north slope of Round Top and has also been collected at Dick's Peak. This similar matted species is distinguishable by its larger, clustered leaves, long leaf hairs that are both simple and forked, and fruits that are often twisted on their stalks. The atypical **Alaska whitlow grass** (*D. albertina*) occurs on semi-moist, subalpine flats and slopes around Mount Rose and Carson Pass. This two- to twelve-inch, erect species has oblanceolate leaves and small yellow flowers that develop into upright fruits, similar to *Arabis*.

Tansy mustard

Tahoe draba

Payson's draba

Payson's draba (*Draba paysonii* var. *treleasei*)

- **1 to 4 inches**
- **early season**
- **dry, rocky slopes and summits**

This species is one of three Tahoe drabas that grow in cushion form with tightly compacted herbage. It has tiny, linear to lanceolate leaves, with long, tangled, branched hairs on both surfaces and margins. It occurs in the north Basin, along the ridges above the Pacific Crest Trail. Elsewhere it may intergrade with the more widely distributed **dense-leaved draba** (*Draba densifolia*), which is common in the north and south Basin, from Castle Peak to Carson Pass. This species has similar linear to oblanceolate leaves that are distinctively sparsely hairy, except for stiff, straight hairs along each leaf margin. **Comb draba** (*D. oligosperma* var. *oligosperma*) has slightly longer, linear basal leaves, each covered with tiny, multiple-branched, comb-like hairs. Less widespread, it is relatively common on the summits of Freel Peak, Jobs Sister, and Mount Rose.

Sierra wallflower (*Erysimum capitatum* ssp. *perenne*)

- 4 to 30 inches
- early season
- dry, open forest, rocky slopes and summits

Tahoe's most widespread mustard family member, Sierra wallflower offers a bright contrast among its varied, earth-tone habitats. It blooms from a single, erect stalk, that ranges considerably in height between low-elevation populations and those persevering above timberline. Wallflower's name comes from European species that are commonly found growing on stone walls.

Miner's pepper (*Lepidium densiflorum* var. *macrocarpum*)

- 10 to 30 inches
- early to mid-season
- dry, disturbed habitats

Lepidium species are distinguishable by their small obovate fruits, which are notched at the top and divided into two chambers carrying one seed apiece. This common, slightly weedy, native species has narrow leaves with toothed margins. *L. virginicum* var. *pubescens*, with slightly larger petals and more prominent, rigid upper stem hairs, is less common, in the far north and south Basin. The European native, **poorman's peppergrass** (*L. campestre*), occurs in wet to moist, low-elevation meadows. It is dis-

tinguishable by its clasping, arrow-head-shaped, upper stem leaves and tiny white flowers.

Western bladderpod (*Lesquerella occidentalis* ssp. *occidentalis*)

- 2 to 6 inches
- early season
- dry, rocky ridges, slopes

This prostrate perennial species occurs above 8,000 feet on the volcanic ridges west of Silver Peak above Shirley Canyon. The grayish green, spoon-shaped leaves are covered with dense, tiny, multiple-branched hairs. This species is distinguished from the drabas by its inflated, elliptic to ovoid fruits borne on stems that mature into an S shape. The Great Basin native *Cusickiella douglasii* has similar inflated fruits, which differ in bearing a single seed. This small, white-flowered species has densely packed basal leaves with straight hairs on the margins. It blooms early in dry sagebrush scrub outside Truckee.

Daggerpod (*Phoenicaulis cheiranthoides*)

- 2 to 8 inches
- early season
- dry, rocky slopes, ledges

Daggerpod occurs in the north and south Basin, typically above 7,500 feet. It is semi-prostrate with grayish green, hairy leaves, ascending to sprawling stems, and small, bril-

Sierra wallflower

Miner's pepper

Western bladderpod

Daggerpod

liant purple flowers that bloom after snowmelt. The distinctive, long, flattened fruits taper gradually to a dagger-like point. The seeds are unwinged and arranged in two rows per fruit chamber. The similar *Anelsonia eurycarpa* has small white flowers that develop into wider, elliptic fruit pods. This species has been found on the summit of Round Top in Carson Pass.

Water cress (*Rorippa nasturtium-aquaticum*)

- 1 to 2 feet
- early to mid-season
- wet habitats

This occasional, low-elevation species has pinnate leaves and small white flowers. It may be distinguished from the similar-looking Brewer's bittercress (*Cardamine breweri* var. *breweri*) by its shorter fruits, which are rounded in cross-section and carry the seeds in two columns. Water cress is a naturalized species from Europe, where it was used to treat iron deficiency.

Western yellow cress (*Rorippa curvisiliqua*)

- 1 to 10 inches
- early to mid-season
- wet to semi-moist muddy flats, slopes

Western yellow cress occurs into the subalpine zone, but is most common on the muddy shores of low-elevation lakes and swamps. It typically is prostrate, with tiny (1 to 2 mm. long) flower petals and deeply, pinnately lobed leaves. The rare **Tahoe water cress** (*Rorippa subum-bellata*) has less pinnately lobed leaves and larger flower petals. It occurs along the sandy shores of southern and eastern Lake Tahoe. Alteration of this habitat for human use over the last century has caused this species to be federally listed as endangered.

Mountain jewelflower (*Streptanthus tortuosus* var. *orbiculatus*)

- 4 to 12 inches
- early to mid-season
- dry forests, rocky slopes, ledges

The elegant mountain jewelflower grows abundantly from lake level into the subalpine zone. The small purple and yellowish-veined flowers reveal an urn-shaped calyx through which four delicate spoon-shaped petals and three pairs of stamens protrude. The fleshy, rounded upper leaves grasp almost around the stem. **Heart-leaved jewelflower** (*Streptanthus cordatus* var. *cordatus*) occurs near the Mt. Rose summit. It has larger, silver-green leaves and pointed sepals with small tufts of hair at the ends. There are twenty-four species of *Streptanthus* in California, many of which grow in marginal soil conditions, including serpentine deposits found throughout the state.

BELLFLOWER FAMILY (CAMPANULACEAE)

This widespread and diverse family, with over 2,000 species worldwide, ranges from small annuals to trees.

Porterella (*Porterella carnosula*)

- 1 to 5 inches
- mid-season
- drying vernal pools

Water cress

Mountain jewelflower

Western yellow cress

Porterella

Porterella is occasional in low-elevation vernal pools from Donner Lake to Hope Valley. The showy, two-lipped flowers bloom close to the ground off short stems. The five stamens are fused. The small alternate leaves are sessile and ovate to triangular. The similar **mountain downingia (*Downingia montana*),** most easily distinguished by flowers that lack individual stems, has been found in the north Basin just outside Truckee. *Heterocodon rariflorum* was collected many years ago in muddy vernal pools near Donner Lake and at Glen Alpine but may no longer occur in Tahoe. This delicate annual species bears tiny blue flowers and alternate, round-cordate leaves that are well spaced along an erect two- to twelve-inch stem.

HONEYSUCKLE FAMILY (CAPRIFOLIACEAE)

Members of this family are shrubs or vines with opposite leaves and a five-lobed calyx and corolla. The ovary is inferior, and there are five stamens. The fruit is typically a berry. The family is small, with only twelve genera, which occur in mostly northern temperate climates. Caprifoliaceae includes the common garden plants honeysuckle (*Lonicera*) and snowberry (*Symphoricarpos*) and the more distantly related elderberry (*Sambucus*), which is occasionally placed in its own family, Sambucaceae.

Double-flowered honeysuckle (*Lonicera conjugialis*)

• 2 to 4 feet
• early to mid-season
• forest openings, moist, rocky slopes

This common, upper montane shrub has elliptic to round leaves and dark red, two-lipped flowers, which are paired on short stems. The upper lip has four shallow lobes, and the lower lip is turned

downward. Three stamens protrude from the upper lip, while the other two are exserted from the center of the flower. The stigma is also exserted. The fruits are clear, semi-translucent, red berries.

Twinberry (*Lonicera involucrata var. involucrata*)

• 4 to 8 feet
• mid-season
• moist stream, lake, meadow margins

Twinberry is occasional in moist habitats, usually below 8,000 feet. The elliptic leaves are strongly veined. The flowers are borne in pairs on a multi-bracted, glandular stem and are tubular to bell-shaped with slightly unequal lobes. Late in the season the bracts turn a deep crimson, surrounding the dark purple fruits. The rare **blue fly honeysuckle** (*Lonicera cauriana*) grows low to the ground (under 2 feet) in wet, grassy habitats such as Sagehen Creek or Osgood Swamp. It has pale yellow flowers with fused ovaries. The leaves are oblong to ovate.

Blue elderberry (*Sambucus mexicana*)

• 5 to 10 feet
• mid- to late season
• semi-moist, open habitats

This occasional, low-elevation species puts forth herbaceous growth that dies back to ground level at the

Double-flowered honeysuckle

Twinberry

Blue elderberry

end of each growing season. Characteristic of the genus, the stems are semi-hollow, filled with a spongy pith. The dark green leaves are pinnate with elliptic, serrated leaflets. The many small, five-lobed, creamy white flowers bunch together in a flat-topped inflores-

cence that may be over a foot wide. The fruits are dark blue berries.

Red elderberry (*Sambucus racemosa* var. *microbotrys*)

- **1 to 4 feet**
- **mid-season**
- **moist, rocky slopes, forest openings**

This variety of red elderberry is abundant from 7,000 up to 10,000 feet, often growing as a semi-matted shrub over the large boulders characteristic of the subalpine zone. (The larger *Sambucus racemosa* var. *racemosa* is common at lower elevations along the coast.) *S. r.* var. *microbotrys* has light green leaves with lanceolate, serrated leaflets and a dome-shaped inflorescence of small white flowers.

Like all elderberries, the plant is toxic, although the bright red berries are edible if cooked. *Sambucus* comes from the Greek word for a musical instrument made from elderberry wood.

Mountain snowberry (*Symphoricarpos rotundifolius* var. *rotundifolius*)

• 2 to 4 feet
• early to mid-season
• forest openings, dry slopes

This medium-sized shrub is common up to 10,000 feet, often on open volcanic slopes. It has pale green, ovate to elliptic deciduous leaves and reddish stems with noticeably shedding bark. The flowers hang in pairs from leaf axils. The small, semi-fused, five-toothed calyx is yellowish, and the narrow, bell-shaped corolla is white to pink. The lower-elevation creeping snowberry (*Symphoricarpos mollis*) is equally common in shady forest understories below 7,500 feet, growing horizontally along vine-like stems. The leaves are entire to occasionally lobed. The fruits of both species are small, whitish berries.

PINK FAMILY (CARYOPHYLLACEAE)

The pink family is generally characterized by opposite, simple leaves, radial, five-petaled flowers, a superior ovary, and five or ten stamens. The calyx is fused in the Silenoideae subfamily, which includes the catchflies, and at least partly free in the Alsinoideae subfamily, which includes the starworts, sandworts, and chickweeds. The common name comes from the deeply lobed petals of some genera, which appear to have been pinked. The scientific name comes from the Latin for clove, a reference to the clove-like scent of some carnations (*Dianthus*). There are eighty-five genera worldwide, with 2,400 species, mostly in higher latitudes of the northern hemisphere.

King's sandwort (*Arenaria kingii* var. *glabrescens*)

• 4 to 12 inches
• early to mid-season
• open, sandy and rocky habitats

The sandworts may be identified within the pink family by their unlobed petals. King's sandwort is abundant in the Basin up to 10,000 feet. It has a strongly branching inflorescence of numerous small white flowers, each with ten exserted, red-anthered stamens. The linear-lanceolate leaves are well spaced up the thin, many-branched, glandular stems. The Great Basin species, needle-leaf sandwort (*Arenaria aculeata*), is common on open to forested, sandy slopes up to 10,000 feet, from Castle Peak southeast along

Red elderberry

Mountain snowberry

King's sandwort

Mouse-ear chickweed

the Carson Range through Marlette Peak. It has tightly bunched, needle-like, mat-forming basal leaves and white flowers that bloom on short pedicels atop six- to ten-inch, erect, leafless stems with small lanceolate bracts at each internode. **Ball-headed sandwort** (*A. congesta* var. *suffrutescens*) has been found in the far north and south Basin. This species has a head-like inflorescence.

Mouse-ear chickweed (*Cerastium fontanum* ssp. *vulgare*)

- **2 to 12 inches**
- **early to mid-season**
- **moist meadows, grassy habitats**

Cerastium species have two-lobed petals, ovaries that have five styles, and cylindrical fruits whose tips have ten distinct teeth. This European native is occasional below 7,000 feet. It has widely spaced, elliptic leaves and inconspicuous white flowers with short petals that barely exceed the sepals. The uncommon **alpine chickweed** (*C. beeringianum* var. *capillare*) occurs above 7,500 feet in moist subalpine habitats in the Mount Rose and Carson Pass areas. It has small, elliptic leaves and white flowers with petals longer than the sepals. The circumboreal **field chickweed** (*C. arvense*) occurs on high-elevation slopes near Mount Rose. It has short-hairy, glandular herbage and relatively large, showy white flowers that bloom atop short (under six-inch) stems. Additional leaf clusters grow out of the lower stem nodes.

Nuttall's sandwort (*Minuartia nuttallii* ssp. *gracilis*)

- **1 to 6 inches**
- **early to mid-season**
- **rocky talus slopes and summits**

This occasional subalpine species has trailing stems, densely packed, awl-shaped stem leaves, and large acute sepals that alternate with five similar-sized, unlobed white petals in a ten-pointed star arrangement. *Minuartia* can be further distinguished by the ovaries, which are divided along only three lines of fissure. The rose to pink-flowered **ruby sand spurrey** (*Spergularia rubra*) occurs as a matted annual or short-lived perennial in moist to dry, typically disturbed habitats below 7,500 feet. The whorled, lanceolate leaves are subtended by whitish, papery bracts. It is native to Europe. **Arctic pearlwort** (*Sagina saginoides*) is uncommon in seepy, subalpine habitats in the north and south Basin. It grows one to three inches tall from a rosette of narrow linear leaves. The tiny, white, hanging flowers do not exceed the green sepals. The ovaries are divided along four to five lines of fissure.

Douglas' catchfly (*Silene douglasii*)

- **4 to 16 inches**
- **mid-season**
- **open, rocky slopes**

Catchflies have a fused, often inflated calyx, deeply lobed petals, and sticky glandular hairs that cover the plant. Such glandular herbage discourages ants and other inefficient, non-flying pollinators. The flowers usually open in the afternoon and wither in the morning sun. Douglas' catchfly is particularly abundant in subalpine, volcanic soils. The two-lobed petals have two linear teeth on the inside, at the base of the petal blade. **Sargent's catchfly** (*Silene sargentii*) occurs on open slopes and summits above 9,000 feet in the south Basin, particularly in the Desolation Wilderness. The outer lobes of the four-lobed petals are much smaller than the inner

Nuttall's sandwort

Douglas' catchfly

Lemmon's catchfly

Lemmon's catchfly (*Silene lemmonii*)

- **6 to 20 inches**
- **mid-season**
- **forest openings**

This forest-dwelling species is common at low to middle elevations in the South Basin. The many nodding flowers have linear, deeply four-lobed petals, white-woolly petal claws, and strongly exserted stamens and styles. Like others in Tahoe, it is named after John G. Lemmon, an amateur who collected many new plants in the northern Sierra during the late nineteenth century. The rare **short-petaled campion** (*Silene invisa*) has been found in the far north and south Basin in open forest and on

middle lobes and distinctly pointed. **Mountain catchfly** (*S. bernardina*) has larger flowers with four equal-sized lobes. This uncommon species occurs in high-elevation, rocky habitats along the southeastern Carson Range.

protected slopes. This four- to six-teen-inch erect species bears several narrow, pink flowers.

Long-stalked starwort (*Stellaria longipes* var. *longipes*)

- 2 to 12 inches
- early to mid-season
- wet to moist meadows, stream banks

This abundant species dots wet, grassy habitats up to 8,500 feet. The five deeply lobed petals resemble a ten-pointed star, punctuated with ten red-anthered stamens and three styles. The flowers perch atop delicate, erect stems with widely spaced, linear-lanceolate leaves. **Straggly starwort** (*Stellaria crispa*) is occasional along damp, shaded meadow edges and stream banks below 8,000 feet. It is prostrate and trailing with evenly spaced, ovate leaves and small flowers composed of five acute sepals and no petals. **Umbellate chickweed** (*S. umbellata*) is uncommon in wet to moist habitats into the subalpine zone. It has tiny flowers arranged in delicate umbels that are borne on slender, branching stems. **Sticky starwort** (*Pseudostellaria jamesiana*) occurs in semi-shady forest, mostly in the south Basin. It has spherical, untoothed, six-valved fruits, and is distinguished by its shallowly lobed petals and glandular herbage.

GOOSEFOOT FAMILY (CHENOPODIACEAE)

This large family is well adapted to saline or alkaline soils. It includes saltbush (*Atriplex* spp.), common in low-elevation deserts, and the pickleweeds (*Salicornia*), abundant in coastal salt marshes. Family members have alternate leaves and small, greenish, wind-pollinated flowers with one to five sepals, no petals, and a superior ovary. Beets, spinach, and chard are in this family, as is the invasive Russian thistle or "tumbleweed" (*Salsola tragus*), which occurs along Tahoe's roadsides.

Pigweed (*Chenopodium album*)

- 6 to 30 inches
- mid- to late season
- disturbed areas

This common European species has small, spherical flowers with five sepals, which generally enclose the flower parts and developing fruit. The flowers bloom in spike-like clusters on erect branches. The lanceolate to deltate leaves have wavy to irregularly toothed margins and are dull green above with small, powdery scales on the underside. The similar **poverty weed** (*Monolepis nuttalliana*) differs in having only three sepals. A third non-native, **Jerusalem oak** (*Chenopodium botrys*), may be distinguished by its pinnate lower leaves

Long-stalked starwort

Orchard morning glory

Pigweed

a common base, and oblong to elliptic, somewhat fleshy leaves. *C. incognitum* has ovate to deltate leaves with untoothed margins.

MORNING GLORY FAMILY (CONVOLVULACEAE)

This primarily tropical family, with five native genera in California, consists mostly of twining and climbing vines. Sweet potato (*Ipomoea batatas*) belongs to this family.

Orchard morning glory (*Convolvulus arvensis*)

• 1 to 3 inches
• mid-season
• moist to dry, disturbed habitats

This invasive European species has large, five-lobed flowers whose petals are fused into a funnel and twisted clockwise in bud. There are five overlapping stamens, a supe-

and glandular herbage. This species is common late in the season along roadsides. Two uncommon native species have been found in Tahoe. *C. desiccatum* occurs in low-elevation moist areas and forest openings in the southwest. It has several stems, which grow from

rior ovary, and a two-lobed stigma. The flowers bloom individually from a main trailing stem. The leaves are shaped like an arrowhead; the two basal leaf lobes flare out perpendicular to the leaf axis, a condition known as *hastate*. This species differs from native morning glories (*Calystegia*) in its bracts, which occur down the stem, well below the small fused sepals.

DOGWOOD FAMILY (CORNACEAE)

This small, diverse, mostly northern temperate family has opposite, simple leaves, flower parts in fours, and an inferior ovary. Dogwoods are California's only representative, with five species.

Creek dogwood (*Cornus sericea* ssp. *sericea*)

• 4 to 10 feet
• early to mid-season
• moist stream, lake edges

Creek dogwood is common into the subalpine zone, where it may grow in dense, shrubby thickets. The stems are dark red to purple. The ovate to elliptic leaves have four to seven pairs of slightly offset, distinct veins that circle out from the main leaf axis. The small flowers form a flat-topped inflorescence. The many grayish blue berries are edible though not particularly sweet. The uncommon **western dogwood** (*Cornus sericea* ssp. *occidentalis*) occurs at low elevations in the southwest Basin. It differs in its larger petals (3 to 4.5 mm. long) and rough-hairy leaf undersurface. The showy foothill species, mountain dogwood (*C. nuttallii*), does not occur in Tahoe. *Cornus* is Latin for horn, a reference to the plants' hard wood.

STONECROP FAMILY (CRASSULACEAE)

This well distributed family occurs in dry temperate regions of the world, especially South Africa. The family is characterized by fleshy, succulent leaves and often colorful flowers. Of California's four native genera, only *Sedum* is present in Tahoe. Stonecrops utilize the Crassulacean Acid Metabolism (CAM) process to flourish in Tahoe's summer-drought climate. Plants open their stomata and take in carbon dioxide at night when moisture stress is low. As temperatures rise the following day, they close their stomata to minimize moisture loss, but are able to continue photosynthesizing by utilizing carbon dioxide stored the night before.

Creek dogwood

Sierra stonecrop

Lance-leaf stonecrop

Lance-leaf stonecrop (*Sedum lanceolatum*)

• 2 to 8 inches
• mid-season
• open volcanic ridges, slopes

This brightly flowered species occurs on dry, subalpine volcanic slopes and plateaus in the far north and south Basin. It has many linear to ovate, often reddish basal leaves and similar stem leaves, which tend to fall off the stem by flowering time. The flowers have

fully opened, lanceolate petals, each with a reddish midrib. The similar **narrow-petaled stonecrop** (*Sedum stenopetalum*) is uncommon on semi-moist rocky soils, ledges, and crevices from mid elevations to 9,000 feet, mostly in the southwest Basin. It differs in its habitat preference, more branching inflorescence, and lanceolate stem leaves that become thin and papery like an onion skin before falling off the stem.

Sierra stonecrop (*Sedum obtusatum* ssp. *obtusatum*)

• 1 to 8 inches
• mid-season
• rocky, sandy ledges

Sierra stonecrop is common in granitic landscapes up to 9,000 feet. The crowded, blue-green to reddish purple basal leaves are round with a slight notch at the tip, while the more widely spaced stem leaves are obtuse. The five greenish sepals are acute, the five white to yellowish petals are obtuse with a slightly pointed tip, and the five stamens are yellow to dark red.

Rosy sedum (*Sedum roseum* ssp. *integrifolium*)

- 2 to 12 inches
- mid-season
- wet to moist cliffs, ledges, rocky streams

This circumboreal species is named for the pink to rose colors of its late-season herbage. It is occasional (at times locally abundant) on wet granitic outcroppings from 7,000 feet through the subalpine zone. The rust-reddish inflorescence consists of numerous, small, four- to five-petaled flowers blooming from prostrate to ascending stems. The basal leaves are in tight rosettes, and the stem leaves are elliptic to ovate with occasional minute serrations. In high latitudes, the native Eskimos include its leaves and shoots as part of their diet.

DODDER FAMILY (CUSCUTACEAE)

Sometimes included with the morning glories (Convolvulaceae), this family consists of only one genus, *Cuscuta,* with about 150 species worldwide, all of which are parasitic, non-photosynthesizing vines. California has eight native dodders.

California dodder (*Cuscuta californica*)

- 1 to 6 inches
- mid- to late season
- open forest, chaparral, grassy habitats

This peculiar annual is occasional into the subalpine zone. It has yellow to orange, thread-like stems, which extract all needed resources from a variety of host plants. The leaves are scale-like. The urn-shaped flowers are white and inconspicuous. Dodder begins life as a seedling that relies on stored reserves until it encounters a nearby photosynthesizing plant. After contact, the seedling shifts its resource demands to the new host. By mid-summer, this parasite may take over an entire area, wrapping up the resident flora in a smothering tangle of orange twine.

SUNDEW FAMILY (DROSERACEAE)

Members of this fascinating family, which includes the Venus flytraps (*Dionaea*), are best known for their consumption of insects as a means to supplement nitrogen intake. *Drosera* is the family's only representative in California, with two species, both of which occur in the Sierra.

Rosy sedum

California dodder

Round-leaved sundew

Round-leaved sundew (*Drosera rotundifolia*)

- **2 to 8 inches**
- **mid-season**
- **wet, boggy habitats**

Sundew is uncommon but locally abundant in wet bogs such as those found at General Creek or Grass Lake. (It was once common in the upper meadows of Squaw Valley, which were paved over for the 1960 Winter Olympics.) This species has tiny, round leaves that spread out in a semi-rosette. The small white flowers bloom from a leafless, semi-erect stem. Each leaf upper surface is covered with long, heavily glandular hairs, which exude a sticky, red secretion that traps small insects. The prey is digested by enzymatic acids secreted by the leaves. Through this extra source of nitrogen, sundews are able to thrive in nutrient-poor soils characteristic of boggy habitats. The rarer **long-leaved sundew** (*Drosera anglica*) differs in its ascending to erect, oblanceolate leaves. Both species can be found at the Mason Bog along west Sagehen Creek.

HEATH FAMILY (ERICACEAE)

This large and diverse family is well represented in the Basin by thirteen genera and many species. Family members are adept at establishing mutualistic mycorrhizal relationships with soil fungi that increase access to essential resources. This has allowed heath plants to occupy resource-poor environments such as granitic sands, shady forest, and acidic wetlands, resulting in a broad diversity ranging from manzanita shrubs to non-photosynthesizing herbs. Heath flowers consist of four to five petals, often fused in an urn shape, and eight to ten stamens surrounding a single style. The urn-shaped flowers are pollinated through a phenomenon known as *buzz pollination*. In this process, bees hang upside down from a flower and vibrate their wings. The anthers respond to the vibration frequency by releasing a limited amount of pollen through narrow apertures. The pollen is deposited on the bee's underside, often attaching to body parts that the bee has difficulty reaching in its normal grooming process. When the bee visits another flower, pollination occurs when the still attached pollen comes into contact with the protruding stigma.

Sugarstick (*Allotropa virgata*)

• 6 to 18 inches
• mid-season
• shady montane forest

Sugarstick is recognizable by the vertical, candy-cane stripes along its flower stalk, particularly striking in the dark habitat in which it grows. Typically rare in Tahoe, it occurs with some frequency during years of high moisture. The small flowers are borne on horizontal axes, where they produce small, dry, wind-dispersed seeds. Sugarstick lacks chlorophyll for photosynthesis, relying instead on soil fungi to obtain nutrients from decaying forest humus.

Green-leaf manzanita (*Arctostaphylos patula*)

• 3 to 6 feet
• early season
• dry open forest, chaparral

There are thirty-eight species of *Arctostaphylos* in California, each with evergreen leaves, white to pink, hanging, urn-like flowers, and peeling, purplish red bark. This species is common at lower elevations in Jeffrey pine forest and montane chaparral. The leaves are large and oval-shaped. Above 7,000 feet, green-leaf manzanita is replaced by **pinemat manzanita** (*A. nevadensis*), a sprawling, mat-forming shrub similar to its larger relative, but with smaller leaves and flowers. It occurs up to 9,000 feet, often associating with huckleberry oak to form a patchwork of shrubby cover across south-facing slopes. Both manzanitas have small, thick, drought-adapted leaves and long taproot systems, which access moisture from deep granitic bedrock fissures. Manza-

Sugarstick

Green-leaf manzanita

Alpine heather

Alpine heather (*Cassiope mertensiana*)

- 4 to 12 inches
- mid- to late season
- moist, rocky seeps

Alpine heather is occasional on northern or eastern exposures in the subalpine zone, occurring abundantly in the Desolation Wilderness. Each small, bell-shaped flower is bordered by five red sepals and nods on its own elongated pedicel. The scale-like leaves are attached to the stem in four concise rows. This plant was described by John Muir as the "fairest and dearest" in the high Sierra.

nitas are named for their small, apple-shaped fruits, used by Native Americans in teas or seasonings. In Tahoe one may encounter black bear scat, rich with partially digested manzanita berries.

Little prince's pine (*Chimaphila menziesii*)

• 4 to 6 inches
• mid-season
• shady forest

This diminutive, semi-woody perennial supplements its photosynthetic production by obtaining resources from decaying humus on the forest floor. Occasional in upper montane forests, it has leathery, dark green, toothed, elliptic leaves and one to three intricate, white to pink, hanging to ascending flowers. Each flower has persistent ovate bracts and protruding stamens that resemble the ridges of a small, royal crown. The alternative name "pipsissewa" comes from a Cree translation of "it breaks into pieces," based on the plant's historical use in treating gallstones. The less common western prince's pine (*Chimaphila umbellata*) has larger, oblanceolate leaves, which are lighter green with toothed margins. The pink to reddish flowers have lanceolate, deciduous bracts and small sepals. Its perceived scarcity in Tahoe is due in part to its tendency to skip blooming altogether in drier years.

Alpine laurel (*Kalmia polifolia* ssp. *microphylla*)

• 4 to 12 inches
• early season
• moist, grassy seeps

This ground-hugging, subalpine subshrub is occasional along Tahoe's west side, particularly in the Desolation Wilderness. The bright rose to pink flowers project out like miniature saucers, with ten stamens reflexed back, each anther tucked into depressions along the edge of the inner corolla. When a bee or other pollinator lands on the flower, the anthers come loose, striking the insect with pollen. The small, evergreen leaves are curled under at the margins. Like all members of this genus, alpine laurel is poisonous to livestock and probably to humans as well.

Labrador tea (*Ledum glandulosum*)

• 3 to 5 feet
• mid-season
• moist lake, stream, meadow margins

Labrador tea is abundant along mid-elevation lake margins, usually forming dense thickets. The clustered flowers form a flat-topped or slightly rounded inflorescence. Each flower has five unfused petals and eight to ten strongly exserted stamens. The alternate leaves have fine yellowish hairs and small glands on the undersurface. Labrador tea gets its name from the bitter beverage brewed from its leaves, the fragrance of which Thoreau described as somewhere between turpentine and strawberries. **Sierra**

Little prince's pine

Labrador tea

Alpine laurel

Sidebells

laurel (*Leucothoe davisiae*) is similar, but with larger, hairless green leaves and an erect, vertical inflorescence with small, hanging white flowers. This northern species makes a rare appearance in Tahoe at Shirley Canyon.

Sidebells (*Orthilia secunda*)

• 4 to 8 inches
• mid-season
• moist, semi-shady forest

Closely related to the prince's pines and wintergreens, this species is occasional from low elevations up to 9,000 feet. It is abundant amidst the lodgepole pines that grow along wet meadows at southeast Donner Lake. Also known as one-sided wintergreen, its bell-shaped flowers grow along one side of the stem, each with a straight protruding style. The flowers and leaves are yellowish green.

Red mountain heather (*Phyllodoce breweri*)

- **6 to 10 inches**
- **mid-season**
- **moist, open to semi-shady habitats**

This common mid- to high-elevation species is especially abundant under the open canopy of forested wetland areas, where it thrives in highly acidic soils. The bright rose-purple flowers are disk-shaped with elongated stamens that stand erect on the fused corolla. The leaves alternate, needle-like, along the stem, giving red heather the appearance of a low, matted conifer.

Pinedrops (Pterospora andromedea)

- **1 to 4 feet**
- **early to mid-season**
- **shady dry forest**

Pinedrops is occasional in Tahoe's upper montane forests, rising through the needled duff like a willowy ghost on a narrow, pink-reddish flower stalk. The leaves are scale-like, and the bell-shaped flowers hang directly from the stem on short pedicels. Like sugarstick and snow plant, pinedrops is a member of the non-photosynthesizing Monotropoideae subfamily. The plant produces numerous, wind-dispersed seeds. (*Pterospora* means winged seed.) The Cheyenne used a boiled concoction of pinedrops to cure nosebleeds. Another monotrop, **fringed pine-sap** (*Pleuricospora fimbriolata*), is extremely rare in low-elevation forests along the west side south to Echo Summit. It has cylindrical, creamy yellow flowers that bloom in a raceme on a single three- to six-inch stem.

White-veined wintergreen (*Pyrola picta*)

- **6 to 12 inches**
- **mid-season**
- **dry, semi-shady forest**

This common upper montane species has dark green basal leaves with whitish borders on the veins. (Younger plants have duller greenish leaves without white borders.) The cream to whitish pink flowers have a distinctly curved, protruding style and hang off an erect, leafless talk. *Pyrola* comes from the pear-shaped leaves of this genus. A seldom seen form, **leafless wintergreen,** occurs in dry, low-elevation forests. Its parallel development to the non-photosynthesizing monotrops illustrates the evolutionary

Red mountain heather

White-veined wintergreen

Pinedrops

tendency of plants in shady environments to dispense with photosynthesis as a source of energy, given the more reliable association with mycorrhizal fungi.

Bog wintergreen (*Pyrola asarifolia* ssp. *asarifolia*)

• 6 to 20 inches
• mid-season
• semi-shady, moist forest, stream banks

This occasional species is most often found under the canopies of lodgepole pines and mountain alders as they encroach upon wetland areas. The nodding, pink to reddish flowers have long, curved, protruding styles. The species name comes from the generic name for wild ginger, whose rounded, green leaves are similar in shape. Wintergreen leaves were historically used to treat bruises; the plants are also known as "shinleaf." Rare in Tahoe, the circumboreal **lesser wintergreen** (*Pyrola* *minor*) has been found in moist habitat at Osgood Swamp. It has smaller leaves, a shorter (under eight-inch) flower stalk, and relatively closed flowers, in which the straight style does not generally protrude past the flower petals.

Snow plant (*Sarcodes sanguinea*)

• 6 to 12 inches
• early season
• shady forest

This highly recognizable, non-photosynthesizing species is a common sight in upper montane forests, pushing through the snow-covered humus layer like an oversized asparagus. The inconspicuous leaves are bract-like, and the many flowers hang from the erect stalk like round, red bells. The genus name means flesh-like in Greek, and the species name means blood-red in Latin. Snow plant is protected from collection under California law.

Sierra bilberry (*Vaccinium caespitosum*)

• 10 to 20 inches
• mid-season
• moist, rocky, meadow or lake edges, seeps

This occasional subshrub is particularly common in the Desolation Wilderness. The thin, oblong leaves have minutely serrated margins. The hanging flowers have a more or less unlobed calyx. Each flower produces a small, bluish berry. **Western blueberry** (*Vaccinium uliginosum* ssp. *occidentale*) is occasional in wet meadows and along lake and stream edges up to 8,500 feet. It has thicker, smooth-margined leaves and an acutely lobed calyx. Typical for *Vaccinium*, these species have deciduous leaves and inferior ovaries. The rare **alpine wintergreen** (*Gaultheria humifusa*) occurs in moist habitats in the Desolation Wilderness. This eight-inch subshrub has small, white, bell-shaped flowers with superior ovaries that form red fruits enclosed by a fleshy calyx. The roundish, evergreen leaves have slightly serrated margins.

PEA FAMILY (FABACEAE)

The large pea family contains such important food crops as peas, beans, peanuts, alfalfa, soybeans, lentils, and clover. Many other Fabaceae are poisonous, however, especially species in *Astragalus* and *Lupinus*. This family is well known for its ability to form mutualistic relationships with nitrogen-fixing bacteria, allowing its members to flourish in nitrogen-poor soils. Tahoe species are all characterized by alternate, compound leaves and a unique five-petal flower structure. The largest petal, the *banner*, projects outward above two *wing*

Bog wintergreen

Snow plant

Sierra bilberry

petals that cover the lower two petals, which form a tightly closed *keel*. Beneath the keel petals are a single pistil and ten stamens whose filaments are often fused. Pollination occurs when a bee lands on the flower's wing petals. The bee's weight causes the petals to open, revealing the flower's reproductive parts, which protrude upward like a fulcrum to brush the bee's dense hairs, depositing and receiving pollen in the process. Smaller insects do not cause the wing petals to spread, and thus pollen is conserved for the preferred pollinator.

Bolander's locoweed (*Astragalus bolanderi*)

- 6 to 16 inches
- mid-season
- open forest, slopes, dry meadow edges

Bolander's locoweed occurs in dry subalpine habitats in the Castle and Ralston Peak areas. It grows sprawling to erect with creamy white flowers that crowd together in groups of seven to eighteen. The fruit grows at an upward angle, curving back toward the stem. **Canada locoweed** (*Astragalus canadensis* var. *brevidens*) occurs at low elevations along roadsides and in other disturbed areas. It grows up to three feet tall with dull cream to slightly purplish flowers that form compact spikes in groups of twenty or more. The two-chambered fruit is stiff and hairy, with a recurved beak. Locoweed is named for its toxic herbage, which damages the optic nerves of grazing animals, causing a distortion of vision that results in unpredictable movements. *Astragalus* is a large genus in California, with at least ninety species.

Whitney's locoweed (*Astragalus whitneyii*)

• 1 to 12 inches
• early season
• rocky ridges, summits

This widespread species is common above 8,500 feet. Like all locoweeds, it has pinnate leaves with an odd number of leaflets. Several varieties occur in Tahoe, one matted form with small pink flowers, and two with white flowers that differ in leaflet size and spacing. All varieties have conspicuous, papery, one-chambered fruit pods, which range in color from yellow-green to pink with burnt-red splotches. The northern Sierra **ball-flowered locoweed** (*Astragalus austiniae*)

occurs in high-elevation, rocky habitats along the western crest south from Castle Peak and in the northern Carson Range to Snow Valley Peak. The small leaves are white-hairy, and the lilac-tinged flowers cluster together in a rounded head, similar to a clover. **Pursh's locoweed** (*A. purshii*) occurs at low elevations just north of Truckee and at high elevations in the Hope Valley-Carson Pass area, a good example of an east-side species exploiting high-desert adaptations to survive in Tahoe's dry, mountainous habitat. This species has large purplish flowers with distinctly recurved banner petals, and dense, white-woolly fruit pods that appear like moth cocoons amidst the gravelly substrate.

Sierra Nevada pea (*Lathyrus nevadensis* var. *nevadensis*)

• 4 to 20 inches
• early to mid-season
• dry forest openings, slopes

Tahoe's only native pea is common at lower elevations in the south and east Basin, typically in dry forest understory where its brilliant flowers enliven the surrounding tones of auburn and green. Pea plants have once-pinnate leaves with an even number of leaflets and twining tendrils that climb the stems of adjacent vegetation, raising the plant toward the sunlight. *Lathyrus* species can be further distin-

Bolander's locoweed

Whitney's locoweed

Sierra Nevada lotus

Sierra Nevada pea

folded, in the bud, and by their flower stigmas, which have hairs on one side (like a toothbrush) instead of on all sides (like a bottlebrush).

Sierra Nevada lotus (*Lotus nevadensis* var. *nevadensis*)

• **2 to 6 inches**
• **early to mid-season**
• **semi-moist, open habitats**

This prostrate, brightly flowered lotus is common at lower elevations in a variety of habitats. *Lotus*

guished from vetch (*Vicia* spp.), an occasional non-native in Tahoe, by their leaflets, which are rolled, not

species generally have miniature palmate leaves with three smooth-margined leaflets and two additional, stipule-like leaflets at the base of the petiole. This species can be most easily recognized by its many small yellow flowers, which turn orange-red after pollination.

Spanish clover (*Lotus purshianus* var. *purshianus*)

- **4 to 16 inches**
- **mid-season**
- **semi-moist to dry, open habitats**

This common species forms densely tangled mats along meadow or stream edges below 7,500 feet. The flowers are cream to pink. The soft, hairy, light green leaves have three ovate leaflets. The small (1.5 to 3 cm. long) fruits are shiny and chocolate-colored. The larger **Torrey's lotus** (*Lotus oblongifolius* var. *oblongifolius*) is occasional in wet, grassy habitats below 7,500 feet. The pinnate leaves have seven to eleven elliptic leaflets. There are one to three leaf-like bracts immediately below the umbel of yellow and white flowers. The non-native **birdsfoot lotus** (*L. corniculatus*) has large, lemon-yellow flowers and five leaflets, the lower two of which are in stipular position. The common name refers to the hanging fruit pods, which were thought to resemble a bird's feet.

Yellow lupine (*Lupinus angustiflorus*)

- **2 to 4 feet**
- **mid-season**
- **dry, semi-open forest, cool slopes**

Lupinus has over 100 recognized taxa in California, at least thirteen of which occur in Tahoe. This species occurs in the north at Sagehen Creek and on protected slopes in the far south. It is distinctive for its deep reddish stems, delicate green leaves with narrow leaflets, and orange-yellow flowers. A hand lens reveals small hairs on the back of the banner petal and a lack of hairs on the upper margin of the keel. Similar plants near Carson Pass with white to pink flowers have been called *L. apertus*, a species also collected near Donner Summit.

Crest lupine (*Lupinus arbustus*)

- **8 to 30 inches**
- **early to mid-season**
- **dry forest openings**

This attractive species is abundant into the subalpine zone. A large white-flowered population covers the lower slopes near Mount Rose, gradually merging with purple-flowered populations south along the Tahoe Rim Trail toward Genoa Peak. The palmately compound leaves, typical of the genus, are green with small straight hairs. It is best identified by the noticeable bump or spur on the end of the

Spanish clover

Yellow lupine

Crest lupine

end of the wing petals toward the tip of the keel.

Tahoe lupine (*Lupinus argenteus* var. *meionanthus*)

• 1 to 3 feet
• mid-season
• open, dry slopes, ridges, forest

This species is common on dry, open terrain above 8,000 feet and in upper Jeffrey pine forest along the Carson Range. It has an upright posture, distinctly silvery leaves, and dull blue, widely spaced flowers with a distinct yellow patch on the banner petal. *Lupinus argenteus* var. *montigenus* is common above 9,000 feet on Mount Rose. It is two feet tall with attractive blue to lavender flowers and greenish gray basal and stem leaves. It has

calyx pointing away from the flower petals and (with a hand lens) the patch of small hairs at the

hairs on the back of its banner petal and a slightly bulging calyx spur. *L. a.* var. *heteranthus* (formerly *L. caudatus*) has a spurred calyx similar to *L. arbustus*, but without the tiny patch of wing petal hairs. This inter-mountain taxon has been collected on dry volcanic slopes and ridges in the Carson Pass area and in the north near Boca Reservoir.

Brewer's lupine (*Lupinus breweri*)

- 1 to 5 inches
- early to mid-season
- dry forest openings, slopes

Brewer's lupine occurs up to 10,500 feet, most commonly in low- to mid-elevation pine forests, where it relies on nitrogen fixation to form fragrant mats of vivid blue and white on the nutrient-poor forest floor. This species has a relatively open inflorescence, hairless upper keel margin, yellow patch on the banner petal, and very small leaflets (less than 1 cm. long). Two varieties, which differ in the presence or absence of hairs on the back of the banner petal, occur in Tahoe.

Green-stipuled lupine (*Lupinus fulcratus*)

- 16 to 40 inches
- mid-season
- dry forest openings

This species is relatively common

into the subalpine zone, usually on granitic soils. The blue and white flowers are widely spaced, with no hairs on the back banner petal. It is best recognized by the two leaf-like stipules at the base of each stem node and the ovate to elliptic leaf lobes. Similar populations that lack leaf-like stipules are sometimes referred to as *Lupinus andersonii*. The degree of taxonomic separation between the two species is still unclear.

Gray's lupine (*Lupinus grayi*)

- 6 to 14 inches
- early to mid-season
- dry forest openings, roadsides

Perhaps the most attractive of Tahoe's lupines, this species is locally common along Old Donner Road in the north and in low-elevation forests around South Shore. The flowers are large and deep blue-purple, often with a distinct yellow spot on the banner petal. Gray's lupine can be identified by its open inflorescence and keel margin, which is densely hairy near the tip and short-hairy near the base.

Torrey's lupine (*Lupinus lepidus*)

- 2 to 24 inches
- early to mid-season
- semi-moist to dry habitats

The widespread Torrey's lupine has a dense inflorescence, persistent

Tahoe lupine

Brewer's lupine

Gray's lupine

Green-stipuled lupine

floral bracts, and hairless banner petal. The keel petal margin is short-hairy near the tip and hair-less toward the base. The smallest variety, *Lupinus lepidus* var. *lobbii*, is prostrate in dry, rocky habitat into the alpine zone, with an inflorescence that does not typically rise above the leaves. The flowers are normally violet-blue with a distinct white banner patch. (This variety has also been referred to as *L. lyallii*). The common *L. l.* var. *sellulus* is larger, with an inflorescence up to a foot tall, well above the mostly basal leaves. *L. l.* var. *con-*

fertus occurs in drying vernal pool and other semi-moist habitats below 8,000 feet. It is common amidst sagebrush scrub near Prosser Reservoir. It is two feet tall, with distinctive lower stem leaves. The northern species, *L. obtusilobus*, occurs on sandy west-facing slopes above Marlette Lake. This one foot, semi-prostrate lupine has a whorled inflorescence that blooms amidst silvery, sparsely hairy stem leaves. There are hairs on the back of the banner petal.

Large-leaf lupine (*Lupinus polyphyllus*)

• **3 to 5 feet**
• **mid-season**
• **moist meadows, stream edges**

This abundant moisture-loving lupine has stout stems, large, dark blue flowers (usually pollinated by bumblebees), and no hairs on either its leaves or keel petal margins. Broad-leaf lupine (*Lupinus latifolius*) is common in low-elevation montane forest openings with some moisture sources, particularly in the southwest Basin. It is distinctive for its height (4 to 7 feet), long, erect inflorescence stalks, and white to pink flowers whose keel petals have small hairs along the first half of the keel margin, but none toward the tip.

White sweetclover (*Melilotus alba*)

• **3 to 6 feet**
• **mid- to late season**
• **roadsides, disturbed areas**

This common European native grows erect along roadsides at low elevations. The yellow-flowered **yellow sweetclover** (*Melilotus officinalis*), another non-native, is slightly less common in similar disturbed habitats. Both species bloom relatively late in the growing season. Sweetclover is named for the sweet fragrance of its foliage.

Shasta clover (*Trifolium kingii var. productum*)

• **4 to 10 inches**
• **early to mid-season**
• **semi-moist, open forest, slopes**

Trifolium has over ten identified species in Tahoe, many of which are non-native. Clovers have palmately compound leaves, usually with three leaflets, which are slightly serrated along the margins. Many have a set of fused bracts (known as an *involucre*) directly below the inflorescence. Shasta clover is occasional in upper montane forests and less frequent on rocky subalpine seep slopes. This delicate plant lacks an involucre, but is recognizable by its drooping inflorescence. The many flowers point downward, while the thin, often forked stem axis projects above the

Torrey's lupine

Large-leaf lupine

White sweetclover

Shasta clover

flower head. *T. gymnocarpon* var. *plummerae* has similar downward-pointing white to pink flowers, which are borne on long, extending pedicels, like an elongated pussypaws. This mat-forming, northeastern species occurs in sagebrush scrub just outside Truckee.

Long-stalked clover (*Trifolium longipes* var. *nevadense*)

• 3 to 10 inches
• early to mid-season
• moist habitats

Long-stalked clover is common in a variety of habitats into the subalpine zone. It lacks an involucre, but can be identified by its leaflets, which are typically long and pointed. The colorful, ball-shaped inflorescence is often two-toned, white and reddish purple.

Red clover (*Trifolium pratense*)

• 4 to 12 inches
• mid-season
• moist to dry habitats, disturbed areas

This common non-native can be identified by its large, reddish inflorescence, lack of an involucre, and faint white chevron on each elliptic leaflet. Red clover is cultivated in Europe for animal feed. Also from Europe, **alsike clover** (*Trifolium hybridum*) has a many-flowered, white to pink, ball-shaped inflorescence and large leaves with ovate leaflets.

White-topped clover (*Trifolium variegatum*)

• 1 to 12 inches
• mid-season
• moist, semi-shady habitats

This diverse species has many California varieties. Several occur in Tahoe, generally in moist, low-elevation environments. The conspicuous involucre is flat and wheel-shaped. The inflorescence consists of one to many red to purple flowers with white or pink tips. The oblong leaflets have rough, serrated margins. In Tahoe, white-topped clover usually forms inch-high mats of few-flowered plants. The diminutive **mountain carpet clover** (*Trifolium monanthum* var. *monanthum*) forms a soft green, two- to four-inch groundcover in semi-shady, seepy areas into the subalpine zone. It has one to three narrowly tubular white flowers (with purple-tipped keels) above an inconspicuous involucre. **Bowl clover** (*T. cyathiferum*) occurs in similar habitats below 8,000 feet. It is named for its involucre, which sits like a bowl beneath many white to greenish yellow flowers with pinkish tips. The involucre margin is wavy and finely toothed.

Long-stalked clover

Red clover

White-topped clover

BEECH FAMILY (FAGACEAE)

This mostly northern, temperate family consists of trees and shrubs with unisexual flowers and alternate leaves. The many male flowers are arranged in catkins or stiff spikes. The inconspicuous, wind-pollinated female flowers usually occur above the male inflorescence. Both types of flowers have minute, five- to six-lobed sepals and no petals. The ovary is inferior. Of the three California genera, oak (*Quercus*) and chinquapin (*Chrysolepis*) occur in Tahoe. The third, tanoak (*Lithocarpus*), can be found at lower elevations in northern coastal and foothill forests. The family also includes the common

North American genera that once occurred in California's prehistoric forests, chestnut (*Castanea*) and beech (*Fagus*).

Sierra chinquapin (*Chrysolepis sempervirens*)

- 2 to 4 feet
- mid- to late season
- dry slopes, forest openings

This common shrub occurs from lake level into the subalpine zone, especially in the south and east Basin as an understory component of Jeffrey pine forest. The evergreen, leathery leaves are hairless and green on the upper surface and golden and scaly on the lower surface. The leaf margins are entire. The male flowers are clustered in stiff, ascending stalks, with the female flowers usually directly below. The distinctive fruits enclose one to three nuts in a spiny, ovoid covering.

Huckleberry oak (*Quercus vaccinifolia*)

- 1 to 4 feet
- early to mid-season
- dry slopes, montane chaparral

Huckleberry oak is abundant on dry southern exposures, usually as a primary member of the montane chaparral community. It has small, leathery, ovate to elliptic, dark green leaves with entire to slightly serrated margins. The male flowers are borne on erect to hanging catkins, while the tiny female flowers bloom in upper leaf axils. The fruit, which takes two years to mature, is an acorn. **Canyon live oak** (*Quercus chrysolepis*) occurs along the lake shore as a large shrub to tree. More common in the foothills, it has oblong leaves with faint lateral, upper surface veins and a golden, hairy (but sometimes hairless) undersurface. *Quercus* has over twenty-five recognized taxa in California. The acorns were a staple in the diet of many Native American tribes.

GENTIAN FAMILY (GENTIANACEAE)

Gentian flowers are usually tubular with a fused calyx and petals. The stamens attach from the petals and alternate between the four or five corolla lobes. The ovary is superior, and the fruit is a two-valved capsule. Many gentians have fringed flower parts, which may aid in depositing pollen on visiting insects. There are seven genera in California, five of which occur in the Basin. In Tahoe, gentian flowers bloom late in the season, undoubtedly contributing to their perceived scarcity.

Sierra chinquapin

Explorer's gentian

Huckleberry oak

Explorer's gentian (*Gentiana calycosa*)

- 6 to 18 inches
- late season
- moist, rocky seeps, slopes

This highly visible perennial is occasional above 7,500 feet, with some low-elevation populations occurring in the southwest Basin. At times locally abundant (such as on Mount Tallac), it may form large patches of deep blue against the late-season landscape. The large flowers have small, white to yellow-green spots (which are thought to act as nectar guides) on the inside of each rounded petal lobe. Between each lobe are thinly forked appendages called *sinuses*. The flowers bloom singly or in small groups on the end of decumbent branches. The opposite leaves are somewhat fleshy and clasp the stem.

Alpine gentian (*Gentiana newberryi*)

- **1 to 4 inches**
- **late season**
- **moist grassy meadows, seeps**

More common in the southern Sierra high country, alpine gentian occurs from low elevations up to 9,500 feet, particularly in the Desolation Wilderness. The large flowers grow close to the ground, arising out of the tangled grass on short stems singly or in groups of two to five. Each flower has a white, vase-like corolla, brown-purple stripes on the outside petals, and acutely fringed sinuses.

Hiker's gentian (*Gentianopsis simplex*)

- **2 to 8 inches**
- **late season**
- **wet to moist meadows**

This annual is occasional, but sometimes locally common, in moist areas below 7,500 feet. The rounded, slightly fringed petals are fused below and free-spreading above, overlapping in a slightly twisting funnel. There are several pairs of elliptic to acute stem leaves. The circumboreal **northern gentian** (*Gentianella amarella*) occurred historically in wet, grassy areas in the southwest Basin, but

has been found recently only at Sagehen Meadows and near Carson Pass. It grows over two feet on an erect, occasionally branched stem with opposite, oblanceolate leaves. The many, small, rose to blue flowers grow in a cyme or out of leaf axils. The brightly colored **charming centaury** (*Centaurium venustum*) occurs in moist habitat between Meyers and Luther Pass off Highway 89. Similar in appearance to Bridge's gilia (*Gilia leptalea*), this annual has opposite leaves, a single stigma, and white spots on the inside throat of its rose-purple flowers.

Green gentian (*Swertia radiata*)

- **2 to 6 feet**
- **mid-season**
- **semi-moist, open slopes**

This striking biennial plant occurs south of Luther Pass and Echo Summit, on open slopes into the subalpine zone. It puts forth a rosette of whorled, oblong leaves with acute tips, then grows to full height and blooms in its second year. The greenish, intricately shaped flowers have spreading sepals, fused only at the base, which alternate with four elliptic, lightly purple-spotted petals. Each petal has a pair of fringed, yellowish green nectar glands and deeply fringed scales between each of the four stamens, which surround the two-lobed stigma. (The flowers

Alpine gentian

Hiker's gentian

Green gentian

give rise to another common name, deer's tongue.)

GERANIUM FAMILY (GERANIACEAE)

Geraniaceae has fourteen genera and over 750 species worldwide. California has two native genera, *Erodium* and *Geranium*, which are also represented in the state by many non-native species. Cultivated plants commonly known as geraniums are in the South African genus *Pelargonium.*

Mountain geranium (*Geranium richardsonii*)

• 1 to 3 feet
• mid-season
• moist, semi-shady forest, seeps

Mountain geranium is occasional in the north and south Basin below 7,500 feet. The palmate leaves have narrow, pointed leaflets. The large flowers have faint purple nectar lines. Each flower has five unfused sepals and petals and ten stamens. The pistil is five-lobed. The five styles are fused to a central axis, forming a long, pointed appendage (the *beak*) in fruit. The five fruit segments coil and uncoil in response to alternating dry and wet conditions, thereby corkscrewing their enclosed seeds into the soil. The name *Geranium* comes from the Greek for crane, an allusion to its long beaked fruit.

GOOSEBERRY FAMILY (GROSSULARIACEAE)

Formerly part of Saxifragaceae, this family consists of the single genus, *Ribes*, which includes the currants and gooseberries, medium-sized shrubs with palmately lobed leaves and small flowers that bloom early in the season, often on a hanging raceme. Each flower is slightly tubular at its base, with five sepals, five petals, and five stamens. The two styles are generally fused at the tip, and the ovary is inferior. The fruit is a berry that, in some species, can be made into jam. There are thirty species of *Ribes* in California, seven of which occur in Tahoe.

Wax currant (*Ribes cereum* var. *cereum*)

- **1 to 6 feet**
- **early season**
- **dry open forests, slopes, rocky summits**

Wax currant occurs from low elevations to the summits of the highest peaks. The small, shallowly lobed leaves have a distinct waxy sheen on the upper surface. The white to pink flowers hang downward in groups of three to seven. The flowers are tubular to bell-shaped, with slightly spreading lobes. The anthers have a cup-like depression at the tip. The fruits are bright red and slightly translucent. High-elevation plants are semi-prostrate, with smaller, glandular-hairy leaves and flowers. **White-stemmed gooseberry** (*Ribes inerme* var. *inerme*) occurs in low-elevation, semi-shady wet meadows and along stream banks. It has branching, pliable stems, thin, deeply lobed leaves, and small flowers with greenish white, reflexed sepals, tiny white petals, and strongly exserted anthers.

Alpine prickly currant (*Ribes montigenum*)

- **2 to 4 feet**
- **early season**
- **semi-moist, rocky habitats, forest openings**

This species is occasional on open slopes and forest from 7,500 feet

Mountain geranium

Wax currant

Alpine prickly currant

into the subalpine. It may be prostrate at higher, windier elevations, on sprawling, spiny stems. The small leaves are deeply lobed and serrated. The flowers are saucer-shaped, with greenish white sepals and rust-red petals. The fruits, which can be deliciously sweet, are

orange-red with glandular bristles. The yellow-flowered **alpine goose-berry** (*Ribes lasianthum*) is occasional from low-elevation forest openings to dry, rocky slopes up to 9,000 feet. It has deeply lobed, toothed leaves that are hairy-glandular, and a slightly spiny stem. The fruits are hairless red berries.

Sierra currant (*Ribes nevadense*)

- **3 to 6 feet**
- **early season**
- **semi-shady forest**

Sierra currant is common from low elevations up to 9,000 feet. It has dense clusters of tubular flowers with pink to reddish, erect sepals and scarcely visible white petals. The leaves are shallowly lobed. The edible berries are dull black-ish blue. Some Tahoe populations (particularly in the east Basin) have an open inflorescence, white to pink flowers, and thin delicate leaves. Others have thicker, heavily veined leaves and dark reddish, clustered flowers.

Sierra gooseberry (*Ribes roezlii* var. *roezlii*)

• 1 to 3 feet
• early season
• dry forest, chaparral openings

This attractive shrub is common below 7,500 feet. The distinctive, hanging flowers have large, reflexed, red-purple sepals, smaller white petals whose margins are curved inward, and exserted red stamens. The leaves are strongly lobed and toothed, and the many branching stems are replete with sharp spines at each leaf internode. The fruits are large, spiny red balls filled with a soft pulp, sweet to the taste for those willing to brave the forebidding exterior.

Sticky currant (*Ribes viscosissimum*)

• 2 to 5 feet
• early to mid-season
• dry, semi-shady forest

Sticky currant is relatively common up to 8,500 feet in forest understories. It has large, heavily veined, gray-green leaves, which are distinctly glandular, and long, white to pink sepals, which spread

at right angles outward from the flower tube. Like wax currant, the anthers of this species have a cup-like depression at the tip.

WATERLEAF FAMILY (HYDROPHYLLACEAE)

Plants in this family typically are perennial herbs with radial flowers blooming on a coiled cyme. The leaves are entire to pinnately lobed. The flowers have five sepals, usually fused at the base, and five free petals. There are five stamens, often exserted, a superior ovary, and two styles with head-like stigmas. The fruit is a two-valved capsule. The family is best developed in western North America. Tahoe has five genera representing a number of species. The common name refers to the watery leaves of certain members of the genus *Hydrophyllum*.

California hesperochiron (*Hesperochiron californicus*)

• 1 to 4 inches
• early season
• open, moist, rocky slopes, meadows

This prostrate species is uncommon up to 9,000 feet in the far north and south Basin, but occurs abundantly in early season at Boca Reservoir and in the meadows around Hope Valley. Each flower has oblong-shaped petals, one

Sierra currant

Sierra gooseberry

Sticky currant

California hesperochiron

style, and two stigmas. The plant has oblanceolate to elliptic, hairy leaves in a basal rosette and funnel-shaped flowers that open singly on short, spreading stems. The similar but rare **dwarf hesperochiron** (*Hesperochiron pumilus*) occurs in moist, rocky areas from low elevations up to 8,000 feet, mostly in the northwest Basin. The small, white, rotate flowers have unfused petals with distinct, purple nectar lines and yellowish bases. The leaves are oblanceolate to oblong and hairless on the underside.

Woolen breeches (*Hydrophyllum capitatum* var. *alpinum*)

- **4 to 8 inches**
- **early season**
- **forest openings**

This early-season plant graces low-elevation slopes and forests soon after snowmelt with lavender flowers that bloom below the deeply lobed leaves. The flowers are bell-shaped and erect with strongly exserted stamens. The species name is Latin for "growing in a dense head," a reference to the plant's rounded inflorescence. It is common in north Basin locations such as Sagehen Creek and Alpine Meadows. **California waterleaf** (*Hydrophyllum occidentale*) occurs in semi-moist upper montane forest in the north Basin, particularly around Castle Peak. (It also can be found to the east near Mount Baldy.) This plant has pinnate leaves with toothed leaflets and white to blue-purple flowers that bloom above the leaves on eight- to twelve-inch stems. *Hydrophyllum* species were historically used in salads by Native Americans.

Purple mat (*Nama lobbii*)

- 4 to 8 inches
- mid-season
- dry, open slopes, gravelly flats

Purple mat is a rhizomatous perennial that forms large, low-lying mats of purple to pink, funnel-shaped flowers that bloom in terminal clusters. This species is rare in Tahoe, but can be found along the highway near Emerald Bay. Like the rest of the plant, the oblanceolate leaves have a sticky, glandular upper surface. The rare **Rothrock's nama** (*Nama rothrockii*) has been found near the summit of Mount Tallac. The sticky, hairy leaves have distinctively *crenate* margins. The pink to purple flowers bloom in a spherical inflorescence atop an eight- to twelve-inch, generally erect stem. **Matted nama** (*N. densum* var. *densum*) is prostrate on a multi-branching, one- to two-inch stem. The white to pale purple flowers bloom from short pedicels amidst narrow, lanceolate, rosetted leaves that are covered with dense, rough, spreading hairs. This Great Basin species has been collected along the southeastern Carson Range.

Sierra nemophila (*Nemophila spatulata*)

- 1 to 10 inches
- early to mid-season
- wet to moist habitats

Nemophilas are small, opportunistic annuals found in a wide range of moist environments. This common upper montane species has white to slightly blue, bowl-shaped flowers with faint bluish veins on the inner petals and an occasional purple spot on the petal lobes. The opposite leaves are oblanceolate, with three to five shallow, usually triangular, lobes. The less common *Nemophila pedunculata* blooms early in wet habitats below 7,000 feet. The leaves are deeply five- to nine-lobed.

Woolen breeches

Purple mat

Sierra nemophila

Timberline phacelia

Timberline phacelia (*Phacelia hastata* ssp. *compacta*)

- **1 to 4 inches**
- **early to mid-season**
- **sandy to gravelly slopes, summits**

Phacelia has ninety-five species in California and many in Tahoe. Phacelias have tubular to bell-shaped flowers with strongly exserted stamens and style and a densely coiled, cyme-inflorescence. This matted subspecies (formerly *P. frigida*) has glandular calyx lobes and spreading hairs. It occurs on high-elevation slopes and ridges into the alpine zone. Below tree-line, it is replaced by **silverleaf phacelia** (*P. h.* ssp. *hastata*), the most common of three erect perennials that occur at lower elevations in a variety of semi-moist to dry habitats. This one- to two-foot species has white to lavender flowers and entire to occasionally lobed basal leaves. The larger **vari-leaf phacelia**

(*P. heterophylla* ssp. *virgata*) has several ascending lateral stems. The basal leaves are *dissected* (deeply lobed, but with remaining "leaf" material along the leaf axis) into triangular segments. **Caterpillar phacelia** (*P. mutabilis*) has a one- to two-foot column-like central stem, strongly coiled, white to greenish yellow flowers, and *compound* basal leaves with three sets of lanceolate to ovate leaflets.

Low phacelia (*Phacelia humilis*)

• **2 to 10 inches**
• **early to mid-season**
• **moist to dry forest openings, brushy slopes, meadow edges**

This showy-flowered annual occurs below 8,500 feet, often forming large purple splotches that spill out from every corner in good moisture years. The elliptic to ovate leaves are borne on an erect stem atop which bloom bell-shaped violet flowers in small coils. The plant is distinctively stiff-hairy and sparsely glandular. **Eisen's phacelia** (*Phacelia eisenii*) is a short (usually under 4 in.), branching species locally common in the Echo Lakes area and scarce elsewhere in the southwest Basin. It has small, white to lavender flowers that bloom in open, uncoiled *cymes* and minutely hairy, elliptic to ovate leaves. The rare *P. racemosa* has been collected at Donner Pass and near Glen Alpine. This

two- to six-inch-tall species has an extremely branching stem with hairless, sometimes glandular, narrowly oblanceolate to linear leaves and small, pale blue flowers. The uncommon *P. quickii* occurs on open, sandy slopes in the southwest and near Mount Rose. It is distinctive for its tightly coiled inflorescence of many small, white to pale blue flowers, which are borne atop a two- to eight-inch, erect stem.

Ballhead phacelia (*Phacelia hydrophylloides*)

• **2 to 10 inches**
• **early to mid-season**
• **dry, semi-shady forest, open slopes**

Ballhead phacelia grows as a prostrate to ascending perennial, usually on conifer-needled forest floors, from low elevations up to 9,500 feet. The oblong to ovate leaves are shallowly to deeply lobed. The small, white to lavender flowers are tightly packed in a spherical inflorescence that rests atop a terminal, leafy stem.

Branching phacelia (*Phacelia ramosissima* var. *eremophila*)

• **1 to 4 feet**
• **mid-season**
• **moist to dry, brushy slopes, forest openings**

This large phacelia is occasional from low elevations up to the al-

Branching phacelia

ST. JOHN'S WORT FAMILY (HYPERICACEAE)

This mostly tropical family is represented in California by one genus, *Hypericum*. The family has opposite leaves and yellow flowers with many exserted stamens, a superior ovary, and a single pistil divided into three distinct style branches. *Hypericum* is Greek for "above a picture," a reference to the way in which plants were hung above religious images to ward off evil spirits during midsummer European festivals. Herbs in Hypericaceae were used in this manner by St. John the Baptist.

St. John's wort (*Hypericum formosum* var. *scouleri*)

- **8 to 20 inches**
- **mid- to late season**
- **moist stream, meadow margins**

This common, low-elevation spe-

Low phacelia

Ballhead phacelia

pine zone, often as part of a diverse montane chaparral community. It is distinctive for its branching, bush-like appearance and pinnate leaves with toothed to compound leaflets. The white to lavender flowers bloom in dense clusters at the ends of long branching stems.

cies bears bright yellow flowers in groups of three to twenty-five on a generally erect, upwardly branched stem. The green sepals are obtuse, sparsely black-dotted, and much smaller than the petals. The many erect, yellow stamens arise out of three main clusters, creating a feathery-topped, golden silhouette. The opposite, clasping leaves are ovate, distinctly veined, and have small black glandular dots near the undersurface margins. The minuscule **tinker's penny** (*Hypericum anagalloides*) is occasional in moist habitats into the subalpine zone. The flowers are similar to its larger cousin, but with sepals equal to or sometimes larger than the petals. Tinker's penny grows as a matted perennial from creeping stolons. The invasive non-native, **Klamath weed** (*H. perforatum*), occurs sporadically in disturbed roadside areas. It has linear to oblong leaves and many flowers (twenty-five to 100) per stem.

MINT FAMILY (LAMIACEAE)

This large, mostly northern temperate family is characterized by square stems, opposite leaves, and irregular, typically two-lipped flowers. The upper lip is entire to two-lobed, and the lower lip is three-lobed. There are four stamens, often in two unequal-sized pairs. The ovary is superior and generally four-lobed with a single style and two stigmas. The fruits consist of four nutlets. The family's natural terpenoids, which defend against herbivory, are fragrant, and the source of many cultivated spices, including lavender, mint, basil, rosemary, sage, and marjoram.

Giant hyssop (*Agastache urticifolia*)

• **2 to 4 feet**
• **mid-season**
• **semi-moist forest openings, slopes**

Common throughout the West, giant hyssop occurs from low elevations to 9,000 feet. It grows on an erect stem with large, upwardly reduced leaves that are lanceolate to triangular with serrated margins. The flowers bloom in mid-summer on many clustered spikes. The unequal stamens and two-lobed style are exserted. The richly fragrant leaves of this genus were used by Native Americans to make a strong tea.

Marsh mint (*Mentha arvensis*)

• **6 to 18 inches**
• **mid-season**
• **wet stream beds, swamp margins**

This circumboreal species is occasional in low-elevation wetland habitats. Its white to slightly pink,

St. John's wort

Giant hyssop

Marsh mint

with toothed, slightly purple margins. The genus is named for the nymph Menthe, who, according to Greek legend, was changed into a plant by the jealous queen of the underworld, Proserpine.

Western pennyroyal
(***Monardella lanceolata***)

• **8 to 20 inches**
• **mid-season**
• **dry, sandy slopes**

Also known as mustang mint, this low-elevation annual has bright pink to rose-purple flowers and an open, candelabra stem structure. It is occasional, and at times locally abundant, in low-elevation sandy habitats in the east Basin from Sand Harbor to Glenbrook.

tubular flowers bloom in clusters at the axils of the leaves and main stems. The leaves are lanceolate

Mountain pennyroyal (*Monardella odoratissima* ssp. *pallida*)

• **8 to 20 inches**
• **early to mid-season**
• **open forests, slopes**

This abundant perennial forms aromatic, rhizomatous patches from low elevations through the subalpine zone, particularly on open volcanic slopes, where it associates with stickseeds, woolly mules ears, and mountain sagebrush. The white to pink flowers are similar to those of marsh mint, but are grouped in dense heads atop erect to ascending stems. The fragrant, ash-green leaves are lanceolate to ovate and often partly folded. Occasional populations with pale lavender flowers and purple-tinged leaves and stems are thought to be the result of hybridization with the eastern Sierra species, *Monardella glauca.*

Self heal (*Prunella vulgaris* var. *lanceolata*)

• **4 to 10 inches**
• **mid-season**
• **moist, semi-shady habitats**

This circumboreal species, with both native and non-native strains in California, is occasional in moist, low-elevation habitats. It grows on a typically erect, single stem with elliptic to lanceolate leaves and a spike inflorescence of densely clustered, white to violet flowers. The upper flower lip is hood-like, and the middle, largest lobe of the lower lip is slightly fringed. The calyx is also two-lipped and pale purplish green. Both the calyx and the bracts below are stiffly hairy. The plant is named for its historical European use as a remedy for various ailments.

California skullcap (*Scutellaria californica*)

• **6 to 16 inches**
• **late season**
• **moist, rocky habitats**

This white- to cream-flowered perennial occurs only in the far north Basin at Shirley Canyon and Granite Chief. It grows erect to ascending from many stems. The leaves are ovate to oblong, becoming sessile upwards. The tubular, two-lipped flowers look like exotic, moonlit spaceships with the recurved lower lip projecting past the hood-like upper lip. **Marsh skullcap** (*Scutellaria galericulata*) has slightly toothed, lanceolate leaves and violet-blue flowers, each with a white-mottled lower lip. This one- to two-foot species was found many years ago in low-elevation marshy areas in the north and south Basin, habitat that has since undergone significant human development.

Western pennyroyal

Mountain pennyroyal

Self heal

California skullcap

Hedgenettle (*Stachys ajugoides* var. *rigida*)

• 8 to 20 inches
• mid- to late season
• drying stream margins, seeps

This abundant perennial herb grows on an erect stem from slender rhizomes throughout the upper montane zone. The ovate to lanceolate leaves are serrated, and the small two-lipped flowers bloom in whorls of six to twelve at upper stem leaf axils. Each flower has scattered violet spots on the inner petals. The leaves, stems, and inflorescence have long, soft, mostly straight hairs that make the plant felt-like to the touch.

Mountain blue curls (*Trichostema oblongum*)

• 1 to 5 inches
• mid-season
• semi-moist, seep habitats

This tiny annual is uncommon below 7,000 feet. The stems and small, widely elliptic leaves are hairy and glandular. The blue-violet flowers form an upward curving, five-lobed tube, the lower lobe in the form of a reflexed lip. The generic name comes from the Greek for hair and stamen, a reference to the long and slender filaments of the flower's exserted stamens. Native Americans used the natural oils of ground-up *Trichostema* plants to stun fish in slow-moving streams or pools.

BLADDERWORT FAMILY (LENTIBULARIACEAE)

The bladderwort family consists of carnivorous herbs found in moist or aquatic environments. It has four genera and 200 species worldwide, mostly in tropical regions.

Common bladderwort (*Utricularia vulgaris*)

• 2 to 12 inches
• mid- to late season
• open standing water

This uncommon species can be locally abundant in marshy habitats such as Grass Lake or Osgood Swamp. The distinctive yellow flowers bloom late in the season on an erect stem that rises above the water surface. The strongly two-lipped corolla resembles flowers in the closely related figwort family. The floating, highly dissected photosynthetic branches (appearing like leaves) contain numerous dark, bladder-like pods along the margins. If minute aquatic insects trigger the sensitive guard hairs, an exterior valve opens, sucking the prey into the bladder chamber with a sudden flush of water. Once inside, the prey is digested by enzymes secreted through glandular hairs on the inner bladder wall. Water is then pumped out through a complex valve system and the trap is reset. A second species, *Utricularia minor*, with smaller flow-

Hedgenettle

Mountain blue curls

Common bladderwort

ers and no bristles along its "leaf" segments, has been found in the middle of Grass Lake.

FLAX FAMILY (LINACEAE)

This mostly temperate family has thirteen genera and over 300 species worldwide. There are three genera in California, only one of which, *Linum,* occurs in the Tahoe Basin.

Mountain flax (*Linum lewisii*)

• 1 to 3 feet
• early to mid-season
• sandy, rocky slopes

This attractive species is common on open slopes in the subalpine zone or above timberline. (It is also a popular ornamental in Tahoe gardens.) It grows on an erect stem with opposite, linear to lanceolate leaves. The blue, radial flowers have five sepals, petals, stamens, and styles. The petals are not fused and fall off shortly after blooming. The fruit is a capsule containing ten seeds. Native Americans used the stems to make strings and cords for snowshoe mesh and fish nets. Linen and linseed oil are made from related European species.

LOASA FAMILY (LOASACEAE)

This family of 200 species in twelve genera includes herbs, shrubs, and even trees in temperate and tropical regions of North and South America. There are three genera in California; only one, *Mentzelia,* occurs in Tahoe. The family has alternate leaves and radial flowers with five petals, five stamens, and an inferior ovary.

Nevada stickleaf (*Mentzelia dispersa*)

• **3 to 20 inches**
• **early to mid-season**
• **dry, open habitats**

This highly variable annual is occasional into the subalpine zone. The rough-hairy, elliptic to ovate leaves range from entire margined to lobed or toothed. The erect stem may be upwardly branched with a varying number of flowers, which bloom singly or in small groups. Each flower has ovate petals, exserted stamens and style, and is subtended by an ovate bract. The

cylindrical fruit develops below the flower, similar to Onagraceae. The eastern Sierra annual, *Mentzelia congesta*, is rare on dry, sandy, shrub-filled slopes in the east Basin up to 9,500 feet. It differs in its dense, head-like inflorescence, which sits atop a receptacle of fused, wedge-shaped bracts, and by its orange-spotted flower petals. The large-flowered blazing star (*M. laevicaulis*), common in the foothills and eastern Sierra, does not generally occur in Tahoe.

MALLOW FAMILY (MALVACEAE)

This large family has sixteen genera in California, but only one in Tahoe. Characteristics include alternate, palmately veined or lobed leaves and many stamens whose filaments are fused into a tube, which is attached to the base of the petals and surrounds the single style. The ovary is superior. The mallow family is known for its *protandrous* flowers, in which the stamens ripen first, then fade to avoid self-pollinating the female parts, which subsequently grow upwards through the fused stamen column. The family includes the crop plant okra (*Abelmoschus*), cotton (*Gossypium*), and the ornamental *Hibiscus*.

Mountain flax

White-veined mallow

Nevada stickleaf

shaped flowers generally bloom one-sided along the tops of many ascending, two-foot stems. The five petals are unfused and heavily veined with white nectar lines.

Bog mallow (*Sidalcea oregana* ssp. *spicata*)

• 1 to 3 feet
• mid- to late season
• wet to moist meadows, seeps

This common upper montane species has palmately lobed basal leaves with crenate margins. The much reduced upper stem leaves are deeply lobed to compound with linear to lobed segments. The rose-pink flowers are borne in a dense spike on an erect, rarely branched stem. The fruit is composed of five to ten wedge-like seg-

White-veined mallow (*Sidalcea glaucescens*)

• 8 to 30 inches
• mid-season
• dry forest edges, open slopes

This common perennial occurs from low elevations to 9,000 feet, most abundantly in the north Basin, where its lavender flowers light up entire mountainsides in midsummer. The palmate leaves are highly variable, becoming deeply five-lobed up the stem. The bowl-

ments, each carrying one seed, which stack side by side in late summer after the last flowers have wilted. A large-flowered form of this variable subspecies, with a denser inflorescence, occurs in the far south Basin at slightly higher elevations.

BUCKBEAN FAMILY (MENYANTHACEAE)

This small family is represented in California by aquatic perennials that root from thick, submerged rhizomes. Formerly considered part of the gentian family, members of Menyanthaceae differ in having alternate leaves.

Buckbean (*Menyanthes trifoliata*)

• 6 to 16 inches
• mid-season
• shallow lakes, marshes, bogs

The circumboreal buckbean is occasional in south Basin marshes such as Lily Lake or Osgood Swamp. It has long-stemmed, lobed basal leaves, which consist of three entire, oblong leaflets. The funnel-shaped flowers bloom raceme-like on an erect one-foot stem. The five white, spreading petals are pink-tinged at the tip with a scaly inner surface, giving the flowers a fringed appearance. The five stamens have thread-like filaments, and the mostly superior

ovary is surrounded by a disk of five glands. The leaves of buckbean were historically used as a substitute for hops in making beer. The roots were ground to make famine bread.

WATER LILY FAMILY (NYMPHAEACEAE)

This primitive family consists of six genera and sixty species of aquatic perennials, many with showy flowers.

Yellow pond-lily (*Nuphar lutea* ssp. *polysepala*)

• 2 to 4 inches
• mid-season
• open water in lakes, ponds

This widespread species occurs in the south Basin below 8,000 feet. It has foot-wide, round to ovate floating leaves and attractive flowers that emerge above the water on long, submerged stems. The flowers consist of seven to nine large, spirally arranged sepals and many, scarcely distinguishable petals and stamens, which surround the numerous stigmas forming a disk-like structure in the flower center. Native Americans roasted the seeds of the berry-like fruits like popcorn or ground them for use in making bread. The non-native **white waterlily** (*Nuphar odorata*) occurs at Lake Lucille and Lake Margery in the Desolation Wilderness. This

Bog mallow

Buckbean

Yellow pond-lily

species has showy, fragrant, white flowers, each with four green sepals. Now placed in Cabombaceae, **watershield** (*Brasenia schreberi*) occurs in similar marshy habitats in the south Basin. It has much smaller, floating oval leaves and solitary reddish purple flowers.

EVENING PRIMROSE FAMILY (ONAGRACEAE)

This showy family has six genera in Tahoe, and some of the Basin's most beautiful wildflowers. The four-petaled flowers are radial and often lobed. Each has four or eight stamens and one pistil, sometimes with a four-lobed stigma. The flowers are differentiated from the four-petaled flowers of the mustard family by their inferior ovary, from which the fruit grows below the fading flower. The fruit is a capsule, and the seeds may be winged with long hairs. This mostly western North American family was named after the wild primroses of the English countryside. The "evening" appendage is a reference to species whose moth-pollinated flowers open at dusk.

Northern sun cup (*Camissonia subacaulis*)

- **1 to 3 inches**
- **early season**
- **semi-moist to dry meadows, slopes**

Northern sun cup is infrequent, but at times locally common, at low to mid elevations in the north and south Basin. It is prostrate with a rosette of large, green lanceolate to elliptic leaves that are entire to slightly lobed. The many medium-sized flowers bloom on short, ascending pedicels. The Great Basin species, *Camissonia pusilla*, occurs on dry, open slopes in the northeast Carson Range. This rare annual has a slender, erect, one- to eight-inch stem, small flowers, and glandular-hairy herbage. It is distinguished from similar-looking *Gayophytum* by its reddish yellow flowers and four-chambered ovaries. *Camissonia* species are distinguishable from those in *Oenothera* by their club-shaped stigmas.

Enchanter's nightshade (*Circaea alpina* ssp. *pacifica*)

- **4 to 8 inches**
- **mid-season**
- **moist, shady stream, meadow edges**

This common upper montane species can be abundant under larger moisture-loving shrubs such as mountain alder. The round, opposite leaves are entire to slightly serrated with an acute tip. Tiny flowers bloom in a raceme on a delicate, erect, glandular stem. Each flower has two white, reflexed sepals, two white, erect petals, two stamens, and a two-lobed stigma.

Diamond clarkia (*Clarkia rhomboidea*)

- **4 to 10 inches**
- **early to mid-season**
- **dry slopes, ridges, open forest**

Common in the western foothills, this species is rare in Tahoe, having been collected many years ago near Fallen Leaf Lake, but found recently only in dry habitats below 8,000 feet in the northwest and at Sagehen Creek. It has four deep-rose, diamond-shaped petals, each with two small lobes at the base, nodding flower buds, and opposite, elliptic leaves. *Clarkia* has over fifty taxa in California. The genus is named after William Clark, the early nineteenth-century leader of the Lewis and Clark expedition.

Northern sun cup

Diamond clarkia

Enchanter's nightshade

Fireweed (*Epilobium angusti-folium* ssp. *circumvagum*)

- **2 to 6 feet**
- **mid- to late season**
- **semi-shady, moist to dry habitats**

Epilobium is a large genus, including a wide variety of perennial

Fireweed

herbs with showy flowers. This pioneer species is abundant up to 9,000 feet, often colonizing burn areas or other disturbed habitats through quickly spreading rhizomes. The alternate leaves are long and linear with a distinctive white mid-vein. The raceme inflorescence is borne on an erect stem. The stigma is strongly exserted and four-lobed. The cylindrical fruits open from the tip along four valves, revealing white, woolly-tufted seeds that disperse in the wind. The less impressive one- to six-foot annual, **parched fireweed** (*E. brachycarpum*), has many small white to rose flowers that bloom off a similar-sized, upwardly branched stem. Like the willow herbs, each flower petal is two-lobed. It occurs in dry, often disturbed habitats below 8,000 feet.

California fuchsia (*Epilobium canum* ssp. *latifolium*)

- **4 to 12 inches**
- **late season**
- **rocky, sandy ledges**

This common species is easy to miss until late season when its multi-flowered blooms cut red swaths across the granitic landscape. Like scarlet gilia, the bright red flowers are long and tubular, which preserves nectar at the flower base for hummingbirds, its preferred pollinator. Each flower has strongly exserted, yellow-anthered stamens and a four-lobed stigma. The grayish green, opposite leaves are linear to ovate. This species was formerly placed in *Zauschneria*.

Sticky willow herb (*Epilobium ciliatum* ssp. *glandulosum*)

- **10 to 40 inches**
- **mid- to late season**
- **wet to moist meadows, stream banks**

Willow herbs have erect to ascending stems, opposite lower leaves, and small white to pink flowers with distinctly lobed petals. Sticky willow herb is one of several difficult to distinguish Tahoe species occurring in upper montane, wetland habitats. Its tall, erect (not ascending) stem bears distinctly veined, lanceolate to ovate leaves and large, glandular, pink to rose-purple flowers. *Epilobium ciliatum* ssp. *ciliatum* has shorter (2 to 6 mm.), white to pink flower petals and a basal leaf rosette. *E. halleanum* grows to twenty inches with short (less than 5 cm.), slightly toothed leaves, small nodding, white to pink flowers, and tiny bulblets on the underground stem and roots. **Oregon willow herb** (*E. oregonense*) blooms early in wet, grassy habitats. It has small, white flowers and rounded to linear, sometimes serrated leaves that are well spaced along four- to twelve-inch, ascending stems. At the stem base are thread-like, aboveground, spreading stems (*stolons*) with minute, rounded leaves.

California fuchsia

Smooth-stemmed willow herb

Sticky willow herb

Smooth-stemmed willow herb (*Epilobium glaberrinum* ssp. *glaberrinum*)

• 1 to 2 feet
• mid- to late season
• moist, rocky slopes, ledges

This species is occasional on rocky, well draining soils into the subal-pine zone. The multiple, ascending stems bear clasping, lanceolate to narrowly ovate leaves and medium-sized, pink to rose-purple flowers. The smooth, hairless herbage is covered with a whitish blue, waxy coating. *Epilobium hornemannii* ssp. *hornemannii* is uncommon in moist, rocky meadows and on stream banks above 7,000 feet, primarily in the Desolation Wilderness. This four- to sixteen-inch species has rose-colored flowers and subtle lines of hairs along its upper stems. *E. lactiflorum* has been found in similar high-elevation habitats. It has ovate to slightly lanceolate leaves, typically white flowers, and longer (2 to 4.5 cm.) fruit stems.

Rock fringe (*Epilobium obcordatum*)

• 2 to 6 inches
• mid- to late season
• moist, rocky slopes, dry subalpine stream beds

This beautiful subalpine species occurs in the north from Mount Rose to Barker Pass and in the south from the Desolation Wilderness to Carson Pass. It grows on a sprawling woody stem that bears large, rose-purple flowers with lobed petals and elliptic to round, bluish green, opposite leaves. The flower stigma is distinctly four-lobed. *Epilobium latifolium* has even larger, rose-purple flowers with unlobed petals, long, pointed sepals, and alternate, elliptic to lanceolate leaves. This rare plant occurs at high elevations in moist, rocky crevices and on talus slopes in the Desolation Wilderness and at Carson Pass.

Gayophytum (*Gayophytum diffusum* var. *parviflorum*)

• 1 to 12 inches
• early to mid-season
• moist to drying, open habitats

Gayophytums are small, hard to distinguish annuals, with tiny, four-petaled, white flowers that open in the morning and fade to pink as the day wears on. Each flower has a two-chambered ovary and eight stamens. The flowers occur individually on delicate stems carrying alternate, lanceolate leaves. The linear, four-valved fruits open like a peeled-banana when the seeds have matured. This species grows in dry forest openings up to 9,000 feet. It has relatively large flower petals (up to 3 mm. long) and slightly knobby fruits, each containing three to eighteen seeds in two overlapping rows. The similar *Gayophytum heterozygum* has narrower, more lumpy fruits containing two to ten ovules, only half of which mature to seed. Of two species with smaller flower petals (under 2 mm. long) that prefer slightly moister conditions, *G. humile* has flat, straight fruits with up to fifty seeds and lateral valves than remain attached at maturity, while *G. racemosum* has fruits with ten to thirty-five seeds and valves that all come free.

Woody-fruited evening primrose (*Oenothera xylocarpa*)

• 1 to 3 inches
• mid-season
• dry, open slopes

This eastern Sierra species occurs on volcanic slopes around Mount Rose from 7,500 to over 9,000 feet. Prostrate, pinnate leaves are dark green with purple splotches. The last, largest leaf segment is round to oblanceolate. Large flowers bloom at dusk and fade to orange-red by mid-afternoon the following day. The flowers have a sweet

Rock fringe

Woody-fruited evening primrose

Gayophytum

odor, which attracts nighttime moth pollinators. Characteristic of the genus, it has eight stamens and a large, four-lobed stigma. **Hooker's evening primrose** (*O. elata* ssp. *hirsutissima*) is occasional along moist roadside seeps at low elevations. It grows on an erect, two- to six-foot stem with alternate, lance-shaped leaves and large, nodding yellow flowers.

BROOMRAPE FAMILY (OROBANCHACEAE)

This family consists of non-photosynthesizing, leafless root parasites. The large, underground stem and root system gives rise to an above-ground inflorescence of generally tubular, two-lipped flowers that resemble those in the related figwort family. The roots are specially adapted to grow from seed and pierce the roots of neighboring plants, absorbing nutrients, organic compounds, and water from the host. The common name refers to the parasitic growths occurring on broom (*Genista*), a common European plant in the pea family. Orobanchaceae has fourteen genera worldwide and two in California.

Corymb broomrape (*Orobanche corymbosa*)

• 1 to 6 inches
• mid-season
• open volcanic slopes, ridges, plateaus

This occasional species can be lo-cally abundant in high-elevation, volcanic scrub habitats in the north and far south Basin. It has many (often greater than twenty) pink to purplish flowers borne on short stems. The linear-triangular calyx lobes are much longer than the unlobed portion of the calyx. The less common **clustered broomrape** (*Orobanche fascicu-lata*) has fewer flowers, longer flower stems (3 to 15 cm.), and short, triangular calyx lobes. Both species prefer hosts in the sage-brush genus *Artemisia*. Clustered broomrape also parasitizes buck-wheat plants. *Orobanche* comes from the Greek words *orobos* (a type of vetch) and *anche*, which means "to strangle."

Gray's broomrape (*Orobanche californica* ssp. *grayana*)

• 1 to 4 inches
• early to mid-season
• moist meadows

This rare species occurs at low el-evations in grassy meadows and along stream banks, where it seems to prefer hosts in the sunflower genera *Aster* and *Erigeron*. The large, white to yellowish flowers have linear-triangular lobes that are longer than the flower tube. **Naked broomrape** (*Orobanche uniflora*) is occasional in rocky seeps and meadows up to 9,000 feet, often in association with its preferred hosts, species of *Sedum* and *Saxifraga*. It has tubular, pink-purple to yellowish flowers with rounded lobes, which are borne singly or in groups of two to three on short stems.

PEONY FAMILY (PAEONIACEAE)

Occasionally included within the buttercup family, Paeoniaceae has a single genus, *Paeonia*, and ap-proximately thirty species world-wide. The peonies are primitive angiosperms, with incomplete dif-ferentiation between leaves and corolla structures and no reduc-tion of floral parts such as stamens and pistils. The family is named after Paeon, physician to the Greek gods. Two species occur in Cali-fornia.

Brown's peony (*Paeonia brownii*)

• 6 to 20 inches
• early season
• dry forest openings, slopes

This intriguing species is especially common in low-elevation, open pine forests. The slightly fleshy leaves are ternately dissected with

Corymb broomrape

Brown's peony

Gray's broomrape

rounded to obtuse segments. The large, drooping flowers consist of five purple-tinged sepals, five to ten rounded, maroon petals with yellowish green borders, and many yellow stamens. The two to five pistils develop into long, slightly curving fruits, which hold several large seeds within their thick exteriors. Native Americans boiled the roots to treat a host of ailments.

POPPY FAMILY (PAPAVERACEAE)

This semi-tropical and temperate family has lobed or dissected leaves, two to four sepals, four or six petals, four to many stamens, and a superior ovary. The name may be derived from the Sumerian *pa pa*, the sound made when chewing poppy seeds. The family includes the Asian species *Papaver somniferum*, from which come opium and its derivatives, heroin, codeine, and morphine. It also includes the lower-elevation state flower, California poppy (*Eschscholzia californica*), whose orange blooms occasionally grace disturbed areas or roadcuts in Tahoe. The large, white-flowered prickly poppy (*Argemone corymbosa*) occurs just miles outside the Basin on the dry, lower, eastern slopes of the Carson Range. Tahoe's two species are in genera that have at times been placed in their own family, Fumariaceae. Flowers in this group tend to be irregular, reserving nectar in elongated petal spurs for long-tongued bee pollinators.

Sierra corydalis (*Corydalis caseana* ssp. *caseana*)

• 1 to 3 feet
• late season
• moist, rocky forest openings

The rare Sierra corydalis has been found near Echo Lakes and along rocky streams near Antone Meadows in the north. A large, almost shrub-like plant, it grows from several to many ascending stems. The leaves are pinnately compound with deeply lobed, rounded leaflets. Each flower has two free petals and two inner petals attached at the tip, forming a fish-like silhouette. This species is on the CNPS Watch List for sensitive and endangered plants.

Steer's head (*Dicentra uniflora*)

• 1 to 4 inches
• early season
• forest openings, rocky habitats

This small, inconspicuous perennial blooms immediately following snowmelt on gravelly flats and slopes from low elevations up to 9,000 feet. Small, dissected, basal leaves have rounded lobes and appear first, parallel to the ground surface. The long-stemmed, drooping flowers bloom shortly thereafter. The outer two petals are free and recurved back, like the horns of a steer, while the two inner petals adhere together at the tip like a snout. The fruits are large ovate capsules containing several seeds. The common Sierra flower, western bleeding heart (*Dicentra formosa*), does not occur in Tahoe.

PHLOX FAMILY (POLEMONIACEAE)

This northern temperate family has nine genera in Tahoe, with many species of perennial herbs and small annuals. The leaves may be opposite (often in whorls) or alternate and entire to pinnately lobed. The typically radial flowers range from relatively open (*Polemonium*) to tube-shaped (*Ipomopsis*). Each flower has five corolla lobes, a narrow, often sheathed calyx, and a superior ovary. There are one style and three stigmas. The five stamens are attached along the insides of the petals. The fruit is a capsule. As a hedge against self-pollination, most phlox flowers are *protandrous*, the style elongating and becoming receptive after the stamens have already dispersed their pollen.

Sierra corydalis

Grand collomia

Steer's head

Grand collomia (*Collomia grandiflora*)

- 8 to 30 inches
- mid-season
- dry, sandy forest openings, disturbed areas

Collomia is best recognized by its alternate leaves and tightly packed, head-like inflorescence of slender, funnel-shaped flowers. The largest of three annuals in Tahoe, grand collomia is occasional and at times locally abundant below 7,500 feet. The salmon-orange, blue-anthered flowers display their distinctive colors with graceful immodesty from a single, upwardly branched, erect stalk. The similar **tiny trumpet** (*Collomia linearis*) has smaller, white to pink flowers and grows in moist to dry meadows on an erect, six- to eighteen-inch stem.

Staining collomia (*Collomia tinctoria*)

• 2 to 8 inches
• early to mid-season
• dry meadows, chaparral and forest openings

This small annual is common in the north Basin below 8,500 feet. It has a typically branching main stem and small, white to pink flowers that bloom sequentially in small clusters of two or more per head. It is named for the yellow stain contained in its roots. The four- to ten-inch **white allophylum** (*Allophylum integrifolium*) differs in its upper leaves, which are palmately three-lobed. It has small, white, tubular flowers with purple-anthered stamens that remain hidden below the corolla lobes. The slightly less common **purple allophylum** (*A. gilioides* ssp. *violaceum*) is shorter, with smaller upper leaf lobes and deep violet flowers.

Bridge's gilia (*Gilia leptalea*)

• 1 to 10 inches
• early to mid-season
• moist to dry, rocky flats, slopes

Gilia differs from *Phlox* and *Linanthus* in its alternate leaves. Bridge's gilia is common up to 9,000 feet, often growing in large, colorful mats. Variable in size, it has delicate stems and leaves and funnel-shaped flowers whose five purple-tipped anthers are well ex-

serted beyond the corolla lobes. Tahoe's two subspecies differ by having purple- or yellow-throated flowers. The tiny **miniature gilia** (*G. capillaris*) grows under four inches tall in moist, grassy and sandy openings. The white, tube-shaped flowers have stamens with bluish anthers.

Scarlet gilia (*Ipomopsis aggregata*)

• 1 to 3 feet
• mid-season
• forest openings, rocky slopes

This erect perennial is common into the subalpine zone, often on open volcanic slopes. It has long, tubular red flowers with flattened, spreading acute lobes. The alternate leaves are pinnate with linear lobes. The flower is pollinated by hummingbirds, whose long beaks probe for nectar deep within the corolla tube, while their breast feathers transfer pollen to the strongly exserted stamens and style.

Globe gilia (*Ipomopsis congesta* ssp. *montana*)

• 1 to 8 inches
• mid-season
• dry, open, rocky ridges, summits

Globe gilia is occasional in exposed, rocky habitats from 8,500 feet into the alpine zone. The small, white, tubular flowers have flat-

Staining collomia

Scarlet gilia

Bridge's gilia

Globe gilia

tened, star-shaped lobes. The flowers cluster in a dense, ball-shaped inflorescence that is prostrate or slightly erect alongside sprawling stems. Its leaves are hairy and palmately three-lobed.

Lavender gilia (*Ipomopsis tenuituba*)

• 1 to 2 feet
• mid-season
• dry volcanic ridges, slopes

Previously considered a subspecies of scarlet gilia, lavender gilia is occasional in high-elevation volcanic habitats in northern and southern Basin locations such as Castle Peak and Carson Pass. This perennial dies after flowering once. The flowers differ from those of scarlet gilia in their pinkish (occasionally white) color and their stamens and style, which are barely, if at all, exserted past the flower tube.

Granite gilia (*Leptodactylon pungens*)

• **4 to 12 inches**
• **mid-season**
• **rocky ledges, crevices**

This rock-loving plant occurs from 7,000 feet into the alpine zone. The densely clustered, alternate leaves are glandular and three to seven palmately lobed with stiff, spiny tips on each leaflet. (The species name means sharp-pointed.) Granite gilia can be distinguished from the similar spreading phlox by its white to pink flowers, which are funnel-shaped (widening from the base in a gradual manner) and twisted in bud.

Whisker brush (*Linanthus ciliatus*)

• **1 to 4 inches**
• **early to mid-season**
• **sandy slopes, dry forest openings**

Linanthus is characterized by opposite, deeply linear-lobed leaves that form spine-tipped whorls along a typically erect stem. This small but showy annual is common up to 8,500 feet, often forming delicate, airy carpets of pink and yellow on open forested slopes. The flowers have narrow tubes with flattened, square-tipped lobes. It is named for the dense, whiskery hairs on its leaf lobes. **Harkness linanthus** (*L. harknessii*) is easy to miss in moist to dry, sandy openings in the upper montane zone. The delicate one- to six-inch stem bears tiny white to pale blue flowers with petals scarcely longer than the narrow sepals.

Nuttall's linanthus (*Linanthus nuttallii* ssp. *pubescens*)

• **6 to 20 inches**
• **early to mid-season**
• **sandy slopes, open forests**

Nuttall's linanthus occurs up to 9,000 feet in the southeastern Carson Range from Genoa Peak to Freel Peak. The showy, white flowers have spoon-shaped petals and barely exserted yellow stamens. This widespread eastern Sierra perennial is named for its densely hairy herbage.

Brewer's navarretia (*Navarretia breweri*)

• **1 to 6 inches**
• **mid-season**
• **open volcanic slopes**

Navarretias are small annuals with opposite, hairy leaves and spine-

Lavender gilia

Whisker brush

Nuttall's linanthus

Granite gilia

tipped bracts that are deeply linear-lobed. The inflorescence is a
spine-bracted head with several to
many small, tubular flowers. This
dry-adapted, Great Basin species
prefers volcanic soils in the north
and south Basin up to 9,000 feet.
The stems are densely branched,

and the yellow flowers have exserted, blue-anthered stamens. The
similar **mountain navarretia** (*Navarretia divaricata* ssp. *divaricata*)
is occasional in dry meadows, flats,
and forest openings up to 8,500
feet. It has small white flowers with
slightly pink-tipped lobes.

Needle navarretia (*Navarretia intertexta* ssp. *propinqua*)

- 1 to 4 inches
- mid-season
- drying vernal pools

One of two white-flowered species found in vernal pool habitats below 6,800 feet, needle navarretia is occasional throughout the Basin, and is best identified by its hairy leaf bases and ovate flower petals. Least navarretia (*Navarretia leucocephala* ssp. *minima*) occurs primarily in the far north Basin. It has long, outwardly exserted bracts and linear-shaped petals.

Spreading phlox (*Phlox diffusa*)

- 2 to 6 inches
- early season
- rocky slopes, summits, open forest

Spreading phlox is abundant from low elevations into the alpine zone. It grows in matted form with densely clustered, linear-lanceolate leaves, which, unlike those of granite gilia, are not spine-tipped. Like all species in this genus (and many in this family), the solitary flowers are *salverform*, meaning they have a narrowly fused tube followed by abruptly spreading corolla lobes. Studies show that pollinated flowers typically change color from white to pink-purple. This transition helps insects to locate unpollinated flowers, while also creating a pastel mosaic for hikers on the sandy, granitic slopes. The similar, white-flowered cushion phlox (*Phlox condensata*) occurs in the alpine zone, spreading over rocks in a tight embrace in response to the abrasive, desiccating winds characteristic of this habitat.

Graceful phlox (*Phlox gracilis*)

- 2 to 8 inches
- early to mid-season
- moist to dry, open habitats

This variable, upper montane annual more closely resembles the collomias and allophylums than its woody-based, perennial cousins. The upright stem may be single to many-branched. The small, white or pink flowers have square-tipped to slightly notched petals. Like all species of *Phlox*, the lanceolate, lower leaves are opposite.

Great polemonium (*Polemonium occidentale*)

- 2 to 3 feet
- mid-season
- wet to moist, semi-shady meadows, stream margins

Polemonium is recognizable within the phlox family by its pinnately compound leaves. This species is locally common in wet to moist habitats from low elevations up to 9,000 feet. It has large, open, blue-purple flowers, each with yellow stamens and a strongly exserted style. The flowers hang off short pedicels from an erect stem.

Brewer's navarretia

Needle navarretia

Spreading phlox

Graceful phlox

Jacob's ladder (*Polemonium californicum*)

- 4 to 12 inches
- early to mid-season
- semi-shady, dry forest

Jacob's ladder is occasional below 9,000 feet, primarily in the north and south Basin. It is particularly common in the open forests around Castle Peak and Carson Pass. The bluish flowers are bell-shaped. The decumbent stems have ladder-like, pinnate leaves, with eleven to twenty-five hairless, lanceolate to ovate leaflets.

Showy polemonium (*Polemonium pulcherrimum* var. *pulcherrimum*)

• 6 to 12 inches
• early to mid-season
• rocky ridgetops, slopes, summits

Perhaps the most attractive of Tahoe's alpine flora, this species grows low on rocky, exposed terrain above 8,500 feet. The medium-sized, light blue to violet flowers bloom in open clusters. The pinnate leaves have small ovate leaflets that overlap one another in fern-like fashion. The spectacular alpine species, sky pilot (*Polemonium eximium*), common throughout the southern Sierra above 10,000 feet, does not occur as far north as Tahoe.

BUCKWHEAT FAMILY (POLYGONACEAE)

The buckwheat family is large, with over twenty genera in California. It is characterized by small flowers that lack true petals, having instead five to six fused sepals, often with a colorful midrib between each sepal lobe. Most genera (not *Eriogonum*) have membranous, sheath-like stipules at the base of each leaf petiole. The ovary is superior, and the single seeds are borne in a dry, three-angled fruit pod. Most species are pollinated by flies or bees. *Rumex* is wind-polli-

nated. The family name translates to "many knee joints," a reference to the swollen stem nodes typical of its members. Buckwheat, sorrel, and rhubarb are common food plants in this family.

Lobb's buckwheat (*Eriogonum lobbii*)

• 1 to 3 inches
• mid- to late season
• rocky ledges, sandy slopes

This genus is one of California's largest, with more than 100 species and 190 recognized taxa, over half of which are endemic to the state. *Eriogonum* species have basal leaves and umbel flower heads, which bloom in clusters atop leafless, but often bracted stems. Each flower has six semi-fused petal-like sepals, usually arranged in two distinct inner and outer whorls, with nine stamens. Lobb's buckwheat is common on mid- to high-elevation, often granitic outcroppings. Its large, round basal leaves and long-stemmed flower heads both lie prostrate. The white to pale yellow flowers turn deep red by late summer. This delicate transformation, typical of many *Eriogonum*, provides some of the most intricate color displays in all of Tahoe. The reddish pigments are thought to allow the plant to conserve heat energy, which in turn fosters seed maturation as daytime temperatures drop.

Great polemonium

Jacob's ladder

Lobb's buckwheat

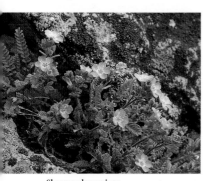

Showy polemonium

Marum-leaved buckwheat (*Eriogonum marifolium*)

- 4 to 16 inches
- mid-season
- sandy to gravelly slopes, open forest

This *dioecious* species forms loose mats on sandy slopes from 7,300 to 9,000 feet. Male plants have dull yellow, small flowers in a head-like inflorescence. Females have larger, brighter flowers in a more open inflorescence. Upper bracts are generally not reflexed, and lower bracts are inconspicuous. Long-stemmed, ovate basal leaves have whitish green, woolly hairs that are gradually shed to reveal an olive-green upper surface. **Hoary buckwheat (*Eriogonum incanum*)** is common in similar habitat into the alpine zone. Also dioecious, it grows under eight inches with pale yellow flowers that bloom above matted, densely clustered elliptic leaves. Leaves remain whitish-woolly on both surfaces throughout the summer. The two species intergrade in the Sierra. *E. microthecum* var. *alpinum* occurs on dry slopes around Hawkins Peak. This low subshrub has white to pink flowers in an open cyme inflorescence and small, oblanceolate leaves that are slightly rolled under and covered with matted, interwoven hairs.

Nude buckwheat (*Eriogonum nudum*)

- 1 to 5 feet
- mid- to late season
- dry, open slopes, forests

Nude buckwheat is abundant into the subalpine zone. The name derives from its tall, leafless stems, which branch upward at various angles, culminating in tight round clusters of small white flowers. The dark green basal leaves are ovate, medium- to long-stemmed, and generally lie flat on the ground.

Butterballs (*Eriogonum ovalifolium* var. *nivale*)

- 2 to 10 inches
- mid-season
- dry, rocky ridges, summits

Butterballs is a cushion- to mat-forming buckwheat found above 8,500 feet. Small, densely packed, oval-shaped leaves are white-hairy. The many white to purple flower umbels rise from the leafy cushion on erect to ascending stalks. The flowers are distinctive for having three round, outer sepal lobes and three inner sepal lobes that are narrower and usually notched. Preferring granitic soils, *Eriogonum ovalifolium* var. *eximium* occurs along the Carson Range from Slide Mountain south to Armstrong Pass. This variety has less compact, slightly larger leaves, with reddish brown margins. The larger, white flowers bloom low to the ground in a more open inflorescence.

Rose buckwheat (*Eriogonum rosense*)

- 1 to 6 inches
- mid-season
- rocky slopes, summits

This Great Basin species is occasional above 8,500 feet along the Carson Range, from Mount Rose (where the plant was first discovered) south to Carson Pass. The small leaves are covered with white-woolly hairs on both surfaces. Rose buckwheat is distinctive for its flowers, whose fused bases lack the stalk-like tube characteristic of the other yellow-flowered buckwheats occurring in Tahoe.

Marum-leaved buckwheat

Butterballs

Nude buckwheat

Rose buckwheat

Spurrey buckwheat (*Eriogonum spergulinum*)

- 6 to 12 inches
- mid-season
- open, sandy slopes, flats

This inconspicuous annual occurs up to 9,000 feet, often forming delicate carpets of tiny (2 mm. wide), white flowers that hover off the ground on multi-branched, reddish, thread-like stems. Each flower lobe has a rose-colored midrib. The whorled basal and upper stem leaves are narrowly linear. The less common eastern Sierra species, **Bailey's buckwheat** (*Eriogonum baileyi*), occurs from Glenbrook south to Luther Pass, in similar habitat. This four- to six-inch species has a less branching stem and flowers borne in a more clustered, sessile inflorescence at stem internodes. The basal leaves are round with matted hairs.

Sulfur buckwheat (*Eriogonum umbellatum*)

• **4 to 20 inches**
• **early to mid-season**
• **open, rocky slopes, chaparral, forest**

This widespread species is abundant into the alpine zone in a variety of dry habitats. Like all yellow-flowered buckwheats in Tahoe (except for *Eriogonum rosense*) the fused flower sepals narrow to a distinct stalk-like tube towards the base. The small, ovate leaves are olive green with whitish hairs below. The bisexual flowers are borne in umbel-like clusters atop leafless stems, subtended by large, leaf-like bracts. Each cluster is composed of many smaller flower umbels subtended by smaller, distinctly reflexed bracts. Of the seventeen varieties occurring in California, most Tahoe specimens have been ascribed to *E. umbellatum* var. *polyanthum*. The rare **Donner Pass buckwheat** (*E. u.* var. *torreyanum*), identifiable by its large, bright green, hairless leaves and striking yellow-orange flower heads, can be found in the northwest near Donner Summit and on the volcanic scree slopes of Silver Peak near Squaw Valley.

Bear buckwheat (*Eriogonum ursinum*)

• **6 to 16 inches**
• **mid-season**
• **sandy ledges, open slopes**

This species is occasional in the northwest Basin, and also in the far south near Carson Pass, in open, sandy habitat up to 9,000 feet. It has thicker, mostly erect flower stems and cream-colored, bisexual flowers with erect upper bracts and exserted yellow-tipped stamens. The light green, woolly leaves are elliptic to ovate and prostrate. Plants tend to grow together in sprawling, one- to two-foot-wide mats.

Wright's buckwheat (*Eriogonum wrightii* var. *subscaposum*)

• **4 to 12 inches**
• **late season**
• **sandy to rocky ledges, outcroppings**

This late bloomer decorates sandy outcroppings below 9,000 feet during August and early September. It grows as a widely matted subshrub with small, elliptic, mostly basal leaves and upwardly branched flower stems, all of which are covered with densely matted, white hairs. Unusual for this genus, the inflorescence is relatively open. The three to many white flowers have green to pink mid-ribs. The strongly exserted stamens are red-anthered.

Spurrey buckwheat

Bear buckwheat

Sulfur buckwheat

Mountain sorrel (*Oxyria digyna*)

- 4 to 16 inches
- mid-season
- rocky cliffs, ledges

The circumboreal mountain sorrel occurs above 8,000 feet in cool, rocky exposures, especially along north-facing ridges between Mount Tallac and Dick's Peak. The rounded, reniform basal leaves turn subtle shades of reddish pink

Wright's buckwheat

over the course of the summer. The small flowers are in a panicle on an erect stem. Each flower has four lobes, six stamens, and two styles. The edible leaves have a strong, sour-lemon taste.

Water smartweed (*Polygonum amphibium*)

• **2 to 6 inches**
• **mid-season**
• **shallow waters of ponds, lakes**

Polygonum is characterized by *cauline* (stem) leaves and five-lobed flowers with three to eight stamens, which bloom in open to dense clusters. Historically common at low elevations, water smartweed is considerably scarcer today due to human alteration of its habitat. The large, oblong to lanceolate leaves float on the water surface, below a slightly emergent, erect stem that terminates in a clustered raceme of many white to pink-rose, tubular flowers. Two intergrading varieties of this circumboreal species exist in Tahoe. The European native **doorweed** (*P. arenastrum*) is common along muddy shorelines and in disturbed areas up to 8,500 feet. Seemingly impervious to trampling, this prostrate species has sprawling stems, linear to elliptic leaves, and small, slightly fused, greenish white to pink flowers that bloom in clusters at the leaf axils.

Bistort (*Polygonum bistortoides*)

• **8 to 24 inches**
• **mid-season**
• **wet to moist meadows**

Bistort is abundant in moist meadows below 10,000 feet. Also known as ladies' thumb for its tightly packed, white-flowered inflorescence, it blooms on an erect, unbranched stem with several oblong to lanceolate, mostly basal leaves. Bistort flowers are pollinated primarily by small beetles and flies that are attracted by the odor of decaying meat. The fragrance resulting from this coevolution has given rise to another common name for the plant, dirty socks. Bistort roots were highly prized as a food source by Native American tribes.

Davis' knotweed (*Polygonum davisiae*)

• **8 to 16 inches**
• **mid-season**
• **dry, open slopes**

This perennial is abundant above 7,000 feet through the subalpine zone, particularly on volcanic mountainsides, where it typically grows in a sprawling fashion following snowmelt. It has light bluish green, elliptic leaves and clusters of small, greenish white flowers. This species is notable for its colorful pink and red, early- and late-season leaves. These colors re-

Mountain sorrel

Water smartweed

Bistort

Davis' knotweed

flect natural pigments that are nor-mally masked by green chloro-phyll, a photosynthetic molecule present in large numbers during the growing season.

Douglas' knotweed (*Polygonum douglasii* ssp. *douglasii*)

• 2 to 20 inches
• mid-season
• moist to dry, grassy meadows

One of several annual knotweeds in Tahoe, this erect species has angled stems with sturdy, elliptic green leaves that are significantly reduced upwards. The inflorescence is open, with small, generally closed flowers that hang from stem internodes. The seeds were collected and ground into meal by Native Americans. The showier *Polygonum douglasii* ssp. *spergulariforme* blooms late in the season in dry habitat near Sagehen Creek. This subspecies has a denser inflorescence of larger, more open, pinkish flowers, which are borne erect off narrower, angled stems. Another meadow species, **dwarf knotweed (*P. polygaloides* ssp. *kelloggii*)**, grows under six inches tall with thin, linear to lanceolate leaves along the entire length of the erect stem. The small white to reddish flowers are sessile at the stem nodes. **Least knotweed (*P. minimum*)** has small white flowers and roundish leaves, which crowd the stem tips. This inconspicuous annual is occasional on semi-moist to dry, sandy slopes in the subalpine zone.

Alpine knotweed (*Polygonum phytolaccifolium*)

• 3 to 8 feet
• mid- to late season
• moist to dry forest openings

Also known as mountain lace, this large plant is common into the subalpine zone in open forest habitats. The small flowers bloom late in the season on erect to hanging panicles, which rise from leaf axils or form the terminal growing shoot of a thick, erect stem. The large, upwardly reduced leaves are lanceolate to ovate with a rounded base and acuminate tip.

Shasta knotweed (*Polygonum shastense*)

• 2 to 10 inches
• mid-season
• sandy to gravelly slopes, forest openings

This inconspicuous perennial grows as a prostrate subshrub in subalpine granitic habitats. The woody stem is gnarled, and the leaves are elliptic to lanceolate. The colorful white to pink flowers are borne in clusters at the upper leaf axils. Each flower has a greenish rib separating the rounded corolla lobes.

Willow dock (*Rumex salicifolius* var. *denticulatus*)

• 1 to 3 feet
• mid-season
• moist to dry habitats

Douglas' knotweed

Alpine knotweed

Shasta knotweed

Willow dock

Rumex has erect stems, alternate, upwardly reduced leaves, and small green, six-lobed flowers that bloom inconspicuously in upright, clustered panicles. After fertilization, the inner three lobes turn reddish and fuse into fin-like membranes. Often, the surfaces between each fin swell into small tubercles. This native species has linear to lanceolate leaves on an upwardly branched stem. The panicle inflorescences are borne on the terminal shoot and from lateral stems down to the lower stem internodes. This variety, without conspicuous tubercles, is the most common of several that occur up to 9,000 feet. The non-native **curly dock** (*Rumex crispus*) grows on an erect stem up to five feet in height with little branching below its terminal inflorescences. Each flower develops three conspicuous tubercles. The margins of the large lanceolate leaves are strongly curled. The non-native **sheep sorrel** (*R. acetosella*) is common in drying, often disturbed areas below 7,500 feet. It grows to sixteen inches on a slender, upwardly branched stem with very small, unisexual flowers that are borne on a narrow raceme. The *hastate* lower leaves resemble tapered arrowheads. The uncommon native **alpine sheep sorrel** (*R. paucifolius*) occurs in moist, grassy areas up to the subalpine zone. This similar species differs by its linear to lanceolate leaves, many stems, and larger, reddish flowers.

PRIMROSE FAMILY (PRIMULACEAE)

This mostly northern temperate family has simple leaves and flower parts in fours or fives. The flowers often bloom from a rising, leafless stem. The ovary is generally superior, and the single style has a head-like stigma. The fruit is a circumscissle or valved capsule. The name comes from the medieval Latin *prima rosa*, which means first rose. There are nine genera in California and two in Tahoe, representing three showy species.

Sierra shooting star (*Dodecatheon jeffreyi*)

- **4 to 16 inches**
- **early season**
- **wet to moist meadows, stream banks, seepy slopes**

Sierra shooting star is common up to 9,000 feet. Its inflorescence is an umbel cluster atop a glandular-hairy, erect stem. The many basal leaves are oblanceolate. The flowers have a short floral tube with strongly reflexed, lavender lobes. The exserted stamens form a partially fused cone through which protrude the female stigma. Like other species with similar floral structure, shooting stars are buzz pollinated. **Alpine shooting star** (*Dodecatheon alpinum*) occurs in similar habitat and differs by its hairless, glandless stems and overall narrower basal leaves. The generic name refers to the hanging flowers, which were thought to resemble a meeting of the twelve principal Greek gods.

Sierra primrose

Sierra shooting star

Sierra primrose (*Primula suffrutescens*)

• 6 to 12 inches
• mid-season
• semi-moist, rocky slopes

This subalpine species is a high country treat, occurring on protected volcanic or metamorphic exposures in the northwest Basin or at Carson Pass. California's only true primrose, it has leathery, spoon-shaped, toothed leaves. The flowers bloom in a single umbel atop a leafless stem. The corolla is funnel-shaped with rich magenta, notched lobes and a yellow throat. Sierra primrose grows as a woody-based, creeping subshrub, typically reproducing asexually through rhizomes. Flower production is highly sensitive to soil moisture; in good years numerous blossoms will turn wide stretches of mountainside bright pink.

PURSLANE FAMILY (PORTULACACEAE)

This family is known for its flowers with only two sepals, which some believe to be derived from modified floral bracts. Purslane leaves are simple, usually linear, and often succulent. The fruit is a capsule.

Pussypaws (*Calyptridium umbellatum*)

• 1 to 6 inches
• early to mid-season
• sandy open flats, slopes

The abundant pussypaws occurs in a variety of open habitats up to 10,000 feet. The basal leaves are oblanceolate to spoon-shaped. The small flowers are clustered in soft, rounded terminal umbels, which

form a basal rosette on long peduncles. As is typical of wide-rang-
ing species, pussypaws varies in
stature, with alpine plants one-
third the size of their low-elevation
kin. Pussypaws regulates internal
temperatures by raising its prostrate flower umbels during the
mid-day heat, then lowering them
back to the still-warm ground surface as air temperatures cool. The
Great Basin species, **rosy pussypaws** (*Calyptridium roseum*), has
been found near the summit of
Snow Valley Peak. This annual
grows under four inches tall with
two-petaled white flowers that
bloom from the axils of spreading
to ascending stems.

Western spring beauty (*Claytonia lanceolata*)

- **2 to 6 inches**
- **early season**
- **moist to dry forest openings, slopes**

Claytonia is known for its variable
number of basal leaves, two opposite stem (*cauline*) leaves, and fruits
with three longitudinal valves that
open from the tip. This species is
occasional between 7,000 and 9,000
feet. The dark green, cauline leaves
are lance-like and sessile against the
erect stem. The several white to
pink flowers have distinct nectar
lines. The uncommon **Sierra claytonia** (*C. nevadensis*) occurs on
high-elevation, seepy-moist slopes

in the vicinity of Freel Peak and
Carson Pass. It has long-stemmed,
elliptic to ovate basal leaves and a
dense, short-stalked inflorescence
with two to eight round-petaled,
white to pink flowers.

Miner's lettuce species are distinctive for their two reniform
leaves, which typically fuse together into a disk around the stem,
above which perch the five to many
small, white to pink flowers. The
succulent stems and leaves were a
favorite of Native Americans and
nineteenth-century miners. *Claytonia rubra* ssp. *rubra* is occasional
below 7,000 feet, typically growing
out of the needle litter at the bases
of large conifers. It has prostrate,
often diamond-shaped basal leaves
and cauline leaf pairs, which may
be only partially or unequally fused
on the stem. Two other closely related species that may occur in
Tahoe can be recognized by their
fully fused cauline leaves and basal
leaves that are either long and linear (*C. parviflora* spp.) or elliptic
at the end of long linear petioles
(*C. perfoliata* spp.)

Long-petaled lewisia (*Lewisia longipetala*)

- **1 to 4 inches**
- **early to mid-season**
- **moist, rocky ledges, slopes**

Lewisias are small perennials
whose fruits open from a cap on
the top. The striking blooms of this

Pussypaws

Western spring beauty

Long-petaled lewisia

rare endemic are a treat for those willing to venture onto the moist, rocky subalpine ledges in the Desolation Wilderness where it occurs. (It also has been found in the north, near Castle Peak and Tinker's Knob.) The long, linear leaves are fleshy, with blunt tips, and grow in a basal rosette. The large white flowers have eight narrow petals and reddish, gland-toothed sepal margins. The equally scarce **pygmy lewisia** (*Lewisia pygmaea*) occurs

in similar habitats in the Desolation Wilderness and Hope Valley-Carson Pass areas. The many flowers have shorter petals (less than 1 cm.) and sepals with jagged margins and rounded tips. This genus is named after Meriwether Lewis, who first encountered *Lewisia* plants on the Lewis and Clark expedition.

Sierra lewisia (*Lewisia nevadensis*)

• **1 to 3 inches**
• **early season**
• **moist, grassy habitats**

This attractive species is common into the subalpine zone. The rosetted, narrowly lanceolate basal leaves are less fleshy than those of its higher-elevation relatives. The flowers bloom singly on short, sometimes elevated, stems. The sepals have generally entire, glandless margins with an acute tip. **Bitterroot** (*Lewisia rediviva*) makes a rare appearance on the high eastern ridges around Marlette Peak. This attractive plant sends shoots directly out of the sandy substrate, upon which bloom large, white to pink flowers that completely cover the withering basal leaves.

Three-leaved lewisia (*Lewisia triphylla*)

- 1 to 6 inches
- early to mid-season
- moist rocky habitats, seeps

This common species up to 9,000 feet is particularly abundant along the trail from Middle Velma Lake to the Rubicon River. It grows on one to several, short, ascending stems that bear clusters of one to ten small, white, hanging flowers. It has two to five semi-succulent, blunt-tipped stem leaves, which are linear and rounded in cross section.

Toad lily (*Montia chamissoi*)

- 1 to 6 inches
- early to mid-season
- wet to moist, grassy habitats

This abundant perennial has opposite, oblanceolate, slightly fleshy leaves and small, open flowers, each with five petals and five pink-anthered stamens that lie flat in a circle against the inner petal margins. In suitable habitat, toad lily may form large colonies that grow from reddish stem runners. The annual **narrow-leaved montia** (*Montia linearis*) blooms earlier in similar habitats below 7,000 feet. It has alternate, fleshy, linear leaves and grows erect on a branched, two- to three-inch stem. The small, half-open flowers hang in groups of three to twelve. The white petals barely exceed the rounded green sepals. **Small-leaved montia** (*M. parviflora*) is uncommon on moist, rocky slopes and cliffs below 9,500 feet. This delicate species grows from thin, rooting stems that form airy mats bearing minuscule, widely spaced, ovate leaves and white flowers that stand out on the steep terrain like solitary beacons.

BUTTERCUP FAMILY (RANUNCULACEAE)

The diverse and attractive buttercup family has nine genera and many showy species in Tahoe. General characters include alternate, often palmately lobed leaves, petal-like sepals, unfused petals, superior ovary, and numerous, spirally arranged stamens and pistils. These latter, "unreduced" floral characteristics indicate that the buttercup family arose at an early stage in the evolution of flowering plants. The family name comes from the Latin for "little frog," a reference to the moist habitats preferred by most family members.

Sierra lewisia

Three-leaved lewisia

Toad lily

Monkshood (*Aconitum columbianum*)

• 2 to 4 feet
• mid-season
• wet, semi-shady habitats

Monkshood is common into the subalpine zone. It is named for its distinctive light blue to deep purple flowers in which the largest of the five petal-like sepals forms a hood over the two petals and enclosed stamens. The flower opening accommodates bumble-

Monkshood

bees, its preferred pollinator. The leaves are palmately lobed with irregularly toothed margins. Occasional plants grow small, deciduous bulblets at stem internodes, which may root to sprout clonal plants the following spring. Highly toxic, this species was also known as wolfsbane from the medieval practice of dipping arrowheads in the plant's juices to kill wolves.

Baneberry (*Actaea rubra*)

- **16 to 32 inches**
- **mid-season**
- **moist, semi-shady forest**

This rare species occurs in deep soils at low elevations. It has one to three ternate leaves with toothed margins and small flowers borne on an ascending raceme. The tiny, white petals and red-tipped sepals are spoon-shaped and half the size of the many conspicuous stamens. The delicate, airy inflorescence glows like an illuminated sphere amidst the surrounding green understory. Baneberry is named for its extremely poisonous fruits.

Drummond's anemone (*Anemone drummondii*)

- **4 to 12 inches**
- **early season**
- **rocky slopes, summits, open forest**

This species is occasional in the northwest (and also at Carson Pass) on rocky habitats in the subalpine zone. (A taller form is rare in forest openings below 8,000 feet.) The large flowers consist of white, petal-like sepals whose undersides are distinctly tinged with blue. The ternate leaves are finely dissected, with linear lobes. The uncommon **western pasque flower**

(*Anemone occidentalis*) can be distinguished by its larger sepals (over 2 cm. long) and plumose styles, which expand out from the fruiting head after pollination like a huge feathery thimble. It occurs at high elevations up to 9,500 feet, primarily along the western Sierra crest. *Anemone* flowers are thought to be the "lilies of the field" described in the New Testament.

Crimson columbine (*Aquilegia formosa*)

- **1 to 3 feet**
- **early to mid-season**
- **moist forest openings, rocky slopes, ledges**

The beautiful crimson columbine is common in a variety of habitats up to almost 10,000 feet. The leaves are two- to three-ternate or deeply three-lobed. The columbine floral structure is noteworthy. Five red, slightly reflexed sepals alternate with five red, tubular, yellow-tipped petals to form a crown-shaped, nodding flower particularly attractive to hummingbirds and bumblebees, which can reach the nectar deep inside the bulbous-tipped petal spurs. (Smaller, short-tongued bees simply chew a hole at the petal spur tip). The cream to pink-flowered, alpine columbine (*Aquilegia pubescens*) does not occur as far north as Lake Tahoe.

Baneberry

Crimson columbine

Marsh marigold

Drummond's anemone

Marsh marigold (*Caltha leptosepala* var. *biflora*)

- 4 to 12 inches
- early season
- wet to moist seeps, stream banks

Often the first sign of life in still soggy subalpine habitats, marsh marigolds create an impressive show at seeps above west Sagehen Creek and Winnemucca Lake. The large, radial flowers have five to eleven white to cream-colored sepals, no petals, and numerous stamens and pistils. The mostly basal leaves are cordate to reniform and occasionally toothed.

Towering larkspur (*Delphinium glaucum*)

- 3 to 8 feet
- mid-season
- moist slopes, stream banks

Towering larkspur is common up to 9,000 feet. The flowers consist of five dark blue sepals, the uppermost forming a long, tapered spur, and four less conspicuous, similarly colored petals. This species has numerous (more than fifty) flowers attached by short pedicels to a single, tall, erect stem. The large, sharply three-lobed, palmate leaves do not occur on the lower twenty percent of the stem. **Mountain larkspur (*Delphinium polycladon*)** has a similar bottle-brush inflorescence on an erect stem, but differs by its shorter stature (two to four feet), more rounded leaflet lobes, less dense inflorescence, and leaves on the lower twenty percent of the stem. This species is rare in moist, rocky habitat in the southwest Basin, occurring more frequently towards Carson Pass. Larkspurs are named for the shape of their flowers, which were thought to resemble the spur of a lark's foot. *Delphinium* comes from the Latin word for dolphin, a reference to the silhouette shape of the larkspur flower buds. All species are highly toxic.

Nuttall's larkspur (*Delphinium nuttallianum*)

- 6 to 18 inches
- early to mid-season
- moist to dry, open slopes, forest, meadow edges

This abundant species has a branching stem, a raceme inflorescence typically consisting of fewer than twelve flowers, and deeply lobed, palmate leaves that occur mostly on the lower stems. As is true for many Tahoe plants, it is named for Thomas Nuttall, a Philadelphia botanist who in 1834 made one of the earliest scientific explorations of the western states. The similar but rare **dwarf larkspur (*Delphinium depaupertum*)** occurs mostly in the southwest Basin, and also at Carson Pass, in semi-moist forest openings and along meadow edges. These species are distinguishable with a hand lens; Nuttall's larkspur seeds have an inflated collar-like shape at their widest end, while dwarf larkspur seeds are bumpy and winged, but lack the inflated collar.

Alpine mouse-tail (*Myosurus apetalus*)

- 1 to 4 inches
- early season
- wet to moist meadows

This easy-to-miss, uncommon annual is locally abundant in the meadows around Meeks Bay and Kings Beach. It has thread-like to

Towering larkspur

Nuttall's larkspur

Alpine mouse-tail

greenish flowers have five to seven spurred sepals and (typically) no petals. Mousetails are named after their long, cone-shaped fruit receptacles, which begin to rise shortly after blooming begins.

Water plantain buttercup (*Ranunculus alismifolius*)

- **4 to 10 inches**
- **early season**
- **wet to moist habitats**

Buttercups are named for the waxy appearance of their yellow flowers, a result of light hitting extra starch molecules in the petals. This abundant low- to mid-elevation species has smooth-margined leaves reminiscent of plants in the water plantain family. The medium-sized flowers have reflexed sepals and

linear basal leaves and a tiny, one-flowered inflorescence that sits atop a short stem. The white to

five to seven petals. Two varieties, which vary in basal leaf size and shape (ovate to lanceolate), occur in the Basin. The diminutive Great Basin species, **desert buttercup** (*Ranunculus cymbalaria*), occurs at low elevations in eastern locations such as Spooner Lake or Boca Reservoir. The ovate to reniform leaves are an inch wide and wavy edged. The small flowers are noticeable for the raised, cylindrical projection of the pistils, which develop into short-beaked fruits. The uncommon **creeping buttercup** (*R. flammula*) blooms along streams or in wet meadows at low elevations. This straggly species has narrow, linear leaves and grows off a thread-like stem, which may take root in the muddy substrate.

Water buttercup (*Ranunculus aquatilus*)

• **1 to 2 inches**
• **mid- to late season**
• **shallow lakes, streams**

This aquatic species occurs at low elevations in bodies of standing or slowly moving water, often forming large, colorful mats by late season. The mostly submerged leaves are finely dissected with thread-like segments, an adaptation that reduces resistance to water flow. (Some populations have "floating" leaves, which are larger and palmately three-lobed.) Due to varia-

tions in water flow, water buttercup flowers may bloom underwater, although their pollinators are typically terrestrial, flying insects.

Alpine buttercup (*Ranunculus eschscholtzii* var. *oxynotus*)

• **2 to 8 inches**
• **early season**
• **wet to moist, rocky slopes**

This large-flowered, prostrate species is found on snow-fed slopes (typically volcanic or metamorphic) from the subalpine into the alpine zones. The open flowers have many petals and numerous raised, greenish pistils, which hover above a halo of stamens. The mostly basal leaves are round to reniform and deeply lobed.

Western buttercup (*Ranunculus occidentalis*)

• **4 to 24 inches**
• **early season**
• **moist, grassy habitats**

Western buttercup is common in Tahoe below 7,500 feet. The deeply three-lobed, occasionally pinnate leaves have toothed leaflets. The clustered fruits have curved beaks. The flowers sit atop thin, ascending stems, often forming tangled masses of yellow and green. **Bird-foot buttercup** (*Ranunculus orthorhynchus* var. *orthorhynchus*) is uncommon at lower elevations. It differs by its deeply lobed, ternate to pinnate leaves, larger flow-

Water plantain buttercup

Alpine buttercup

Water buttercup

Western buttercup

ers, and longer (2 to 4 mm.), straighter fruit beaks. Also uncommon, **woodland buttercup** (*R. uncinatus*) has smaller flowers with petals that barely exceed the sepals. The fruit beaks are strongly curved, and the leaves are three-lobed with toothed leaflets.

Fendler's meadow rue
(*Thalictrum fendleri* var.
fendleri)

• **2 to 6 feet**
• **early to mid-season**
• **moist, semi-shady forest,
  streams**

Meadow rue is abundant from low

elevations into the subalpine zone. Leaves have rounded, slightly toothed leaflet segments. This species is *dioecious*, with male and female plants. Male flowers have four to five greenish white sepals, no petals, and numerous white to purplish stamens, which hang like chimes, dispersing pollen along the occasional gusts of wind in the forest understory. Female blossoms sit upright, with many small, curving pistils that reach out like the tentacles of an octopus. **Few-flowered meadow rue** (*Thalictrum sparsiflorum*) is occasional in wet, shady areas in the upper montane zone. This species has bisexual flowers with more conspicuous, petal-like sepals.

BUCKTHORN FAMILY (RHAMNACEAE)

This family consists of shrubs or small trees with simple, alternate leaves and small, radial flowers with four or five petals and stamens and a single inferior ovary. Four of the six California genera are native to desert scrub communities along the Mexican border. The other two, *Ceanothus* and *Rhamnus*, form an integral part of the state's chaparral communities.

Mountain whitethorn (*Ceanothus cordulatus*)

- **2 to 4 feet**
- **early to mid-season**
- **semi-dry, open forest, slopes**

Mountain whitethorn is abundant in the upper montane zone. It has small, ovate to elliptic, gray-green leaves that alternate along semirigid, thorny, grayish white stems. It is also known as snowbush for its fluffy clusters of white flowers. There are forty-three species of *Ceanothus* in California, many in cultivation as ornamentals. Sometimes referred to as California lilac, the genus typically has five similarly colored sepals and petals, five stamens, and a superior three-lobed ovary. The fruits are three-valved capsules, which typically open with an audible pop on hot afternoons, dispersing seeds and leaving behind a three-ridged indentation resembling a peace sign.

Mahala mat (*Ceanothus prostratus*)

- **2 to 8 inches**
- **early season**
- **dry, open forest**

This mostly prostrate subshrub is common in the north (and occasional in the southwest) in the upper montane zone, particularly in Jeffrey pine forest. The dark green, obovate leaves are spiny-margined and firm. It grows from widely spreading branches, forming roll-

Fendler's meadow rue

Mahala mat

Mountain whitethorn

Tobacco brush (*Ceanothus velutinus* var. *velutinus*)

- **3 to 6 feet**
- **early to mid-season**
- **dry, open forest, slopes**

Tobacco brush is common in the upper montane zone, especially along the trail above Eagle Lake in the southwest. It has flexible, branching stems, which are brown and spineless. The large green leaves are slightly shiny on the upper surface, glaucous-silky below, and minutely gland-toothed along the margins. The white-petaled flowers have yellow stamens and are borne in a large, open panicle.

ing carpets across the forest floor. The small, blue to purple flowers each have five exserted, yellow stamens.

Sierra coffeeberry (*Rhamnus rubra*)

- 3 to 7 feet
- mid-season
- moist, rocky slopes, stream edges

This occasional upper montane species has distinctive elliptic leaves that are shiny green with faint venation and have entire to slightly toothed margins. The ascending stems have reddish to gray bark and twigs. The small flowers bloom in umbel clusters of six to fifteen with five white to greenish yellow petals and sepals and a non-exserted style. The black fruit is round with two stone-like seeds, resembling a small coffee bean. The northern species, **alder-leaf coffeeberry** (*Rhamnus alnifolia*) is rare in the north Basin in wet forest and meadow edges within a limited elevation range centered around 6,800 feet. It has grayish bark, brown twigs, and small petalless flowers that bloom singly or in clusters of two to three. The black fruits have three seeds.

ROSE FAMILY (ROSACEAE)

The rose family consists of trees to annual herbs with typically alternate leaves and radial flowers. Each flower generally has five sepals, five unfused petals, five to many exserted stamens, one to many pistils, and a saucer-shaped receptacle known as the *hypanthium*. The fruits vary considerably, from the many-seeded strawberry to the one-seeded plum or feathery achene characteristic of mountain mahogany. The family includes such important food crops as apples, almonds, pears, cherries, peaches, raspberries, and strawberries, as well as ornamentals such as roses and *Spiraea*.

Serviceberry (*Amelanchier utahensis*)

- 5 to 10 feet
- early season
- moist to dry, semi-open habitats

This attractive shrub is common at lower elevations and occasional in semi-moist, rocky habitat in the subalpine zone. The soft, dark green leaves are flat and slightly serrated. Each flower has large, wavy petals, which appear like giant, white snowflakes against the subdued terrain. The bluish black, slightly sweet fruits were a favorite of Native Americans. **Glabrous serviceberry** (*Amelanchier alnifolia var. pumila*) differs by its hairless upper leaf surface and five (as opposed to two to four) styles. It is less common in moister areas in the south Basin, at times growing in tree form, with thick trunks and large, round leaves, along the southeastern lake shore.

Tobacco brush

Serviceberry

Sierra coffeeberry

Curl-leaf mountain mahogany (*Cercocarpus ledifolius*)

- 6 to 12 feet
- mid-season
- dry, open slopes

Curl-leaf mountain mahogany

This east-side species has a sporadic distribution, with individuals occurring in the southwest and along the southeastern lake shore, and larger populations abundant on dry slopes in the far north and south. It grows as a large shrub, with small, leathery, elliptic leaves that curl inward at the margins. The inconspicuous, tubular flowers lack petals. Showy displays of feathery fruits drape the entire plant in a silvery sheen later in the growing season. The genus name means tailed fruit in Greek.

Mountain strawberry (*Fragaria virginiana*)

- 1 to 4 inches
- early season
- semi-moist forest openings

This widespread North American species is common at low elevations, often forming large colonies off creeping horizontal stems that root to form new plants. The soft, light green leaves are ternate with toothed leaflets. The large, white flowers bloom early and low to the ground. Each produces a many-seeded strawberry fruit, which is tasty to the local fauna.

Big-leaf avens (*Geum macrophyllum*)

- 1 to 3 feet
- early to mid-season
- semi-shady, moist meadows, stream margins

This moisture-loving plant, common in the foothills, is occasional below 7,500 feet. It has clasping, pinnate leaves with large (up to 4 inch long), cordate to reniform, toothed, terminal leaflets. The yellow flowers, each with a raised, greenish, button-like collection of pistils in the middle, nod horizontally on erect to leaning stems. The fruiting head is spherical, with many tiny hooked styles that protrude outward long after the flower petals have fallen away.

Prairie smoke (*Geum triflorum*)

- 6 to 20 inches
- early to mid-season
- open slopes, plateaus, ridges

This species is occasional in the subalpine volcanic or metamorphic habitats that occur in the northern and far southern Basin. Also known as purple avens, its drooping flowers are pink to purple-tinged, with vase-shaped sepals and petals. At the flower base are five similarly colored, outstretched bractlets. After fertilization, the petals and sepals fall away, leaving behind numerous achenes attached to long, plumose styles that flutter in the wind, giving rise to another common name for the plant, old man's whiskers.

Creambush (*Holodiscus discolor*)

- 2 to 5 feet
- mid-season
- open, rocky slopes

A common member of the montane chaparral community up to 9,500 feet, creambush bears numerous, saucer-shaped flowers that bloom in panicles at the ends of mostly leafless stems. This species has elliptic upper stem leaves, which are toothed above the middle of the leaf margins. The similar cliffspray (*Holodiscus microphyllus* var. *microphyllus*), also known as rock spiraea, occurs on rocky outcroppings, usually at

Mountain strawberry

Prairie smoke

Big-leaf avens

Creambush

higher elevations. This species is distinguished by its shorter (under 2 cm.), obovate upper stem leaves, which tend to mix with flowers higher on the stem. In Tahoe, these character distinctions may blur as a result of hybridization.

Dusky horkelia (*Horkelia fusca* ssp. *parviflora*)

- 1 to 2 feet
- early to mid-season
- forest openings, meadow edges

Horkelia is a large genus in California, with over twenty-five recognized taxa. This species occurs sporadically in the upper montane zone. It has a distinct gray-greenish appearance and pinnate leaves with an odd number of wedge-shaped, toothed leaflets. The small, white flowers have wedge- to triangular-shaped petals and reddish sepals and bracts. The mostly leafless, reddish stems are glandular-hairy.

Gordon's ivesia (*Ivesia gordonii*)

- 2 to 10 inches
- early to mid-season
- dry, rocky ridges, summits

Gordon's ivesia is the most common of several high-elevation, yellow-flowered *Ivesia* species occurring in Tahoe. It has green, basal, pinnate leaves with ten to sixteen leaflets per side, each with from four to eight lobes. The yellow, star-like flowers have five stamens, two to four pistils, and oblanceolate petals. The rare **club moss ivesia** (*I. lycopodioides*) has thirty-five leaflets per side, obovate petals, and five to fifteen pistils. It has been found on high-elevation talus slopes near Freel Peak and at Carson Pass.

Mat ivesia (*Ivesia shockleyi* var. *shockleyi*)

- 1 to 6 inches
- early to mid-season
- dry, rocky ridges, summits

This rare species occurs on wind-swept volcanic ridges near the Mount Rose summit and in the vicinity of Tinker's Knob. It has a matted growth form and crowded leaves with five- to ten-lobed leaflets per side. Each small flower has five stamens. The southern Sierra dwarf ivesia (*Ivesia pygmaea*), distinguishable by its ten stamens and numerous pistils, does not occur as far north as Tahoe.

Mousetail ivesia (*Ivesia santolinoides*)

- 6 to 30 inches
- mid-season
- open, sandy flats, slopes

This occasional, often locally common, white-flowered species occurs from low elevations into the subalpine zone, especially on granitic soils, in the south Basin. The distinctive silvery, mousetail-like leaves have sixty to eighty tiny, barely distinguishable leaflets per side. The numerous white flowers bloom atop long, ascending, often multiple-branched stems, adding an airy third dimension to the splotchy granitic terrain. Each flower has rounded petals and fifteen stamens.

Dusky horkelia

Mat ivesia

Gordon's ivesia

Mousetail ivesia

Plumas ivesia (*Ivesia sericoleuca*)

- 4 to 20 inches
- mid-season
- moist to dry, rocky meadows

Plumas ivesia occurs in drying meadows at the far north, from Truckee to Stampede Reservoir. It has prostrate, green leaves with twenty to thirty-five leaflets per side and good-sized, white flowers with paddle-shaped petals and around twenty stamens. Endemic to California, Plumas ivesia is on the CNPS list of rare and endangered plants.

Fan-leaf cinquefoil (*Potentilla flabellifolia*)

- 4 to 12 inches
- mid-season
- moist meadow, stream, lake edges

Fan-leaf cinquefoil is one of three *Potentilla* found primarily at high elevations in Tahoe. It has five deep yellow petals that alternate with large, persistent, green sepals (compare to Tahoe's buttercups), twenty stamens, and numerous pistils with deciduous styles. It can be recognized by its green, ternate leaves with three fan-shaped, toothed leaflets, but is perhaps best appreciated for its large, enticing flowers. **Drummond's cinquefoil** (*P. drummondii* var. *breweri*) grows semi-prostrate in moist to drying, open subalpine terrain. It has whitish green, pinnate leaves with overlapping, palmately lobed leaflets. *P. d.* var. *drummondii* occurs in lower-elevation, moist grassy areas (such as Paige Meadows) and has green, pinnate leaves with two to four generally separated, toothed leaflets per side. Cinquefoils are named after the French term for five leaflets, a characteristic more typical of European species.

Shrubby cinquefoil (*Potentilla fruticosa*)

- 1 to 3 feet
- early to mid-season
- open, rocky slopes, ridges, and plateaus

This attractive, largely subalpine species is occasional in open, rocky (usually volcanic) habitats. It is the only *Potentilla* species that grows in shrub form in California. Leaves are pinnate, with two to three narrow, elliptic leaflets per side. Its large, showy, yellow flowers make this a favorite of native plant gardeners and horticulturists.

Sticky cinquefoil (*Potentilla glandulosa* ssp. *ashlandica*)

- 4 to 30 inches
- early to mid-season
- moist to dry slopes, open forest

Sticky cinquefoil is abundant up to 9,500 feet. This variety has pale yellow flowers and pinnate leaves. The leaf base (where the leaf attaches to the main stem) is hairless, and each leaflet has ten teeth. Other subspecies in Tahoe include *Potentilla glandulosa* ssp. *reflexa*, with small petals shorter than or equal to its sepals, and the cream-colored *P. g.* ssp. *nevadensis*, common in the Carson Pass area. **Multi-leaved cinquefoil** (*P. millefolia*) is a Great Basin species that occurs in west Sagehen Meadows in moist, open grassy habitat. It is prostrate with pinnate basal leaves

Plumas ivesia

Fan-leaf cinquefoil

Shrubby cinquefoil

Sticky cinquefoil

consisting of five to thirteen many-lobed leaflets. Another east-side species, **biennial cinquefoil** (*P. biennis*), occurs along the moist, grassy, muddy shores of Spooner Lake. It has ternate leaves and flower petals smaller than the sepals.

Graceful cinquefoil (*Potentilla gracilis* var. *fastigiata*)

• 1 to 3 feet
• mid-season
• moist meadow edges, forest openings

Graceful cinquefoil is abundant in a variety of semi-moist habitats up to 9,000 feet. It has distinctive large palmate leaves with five to seven toothed leaflets and ascending, many-branched stems. Tahoe's variety is one of four occurring in California. More typical in the southern Sierra, *Potentilla diversifolia* var. *diversifolia* has been found above 8,000 feet in the Desolation Wilderness. This four- to sixteen-inch species differs by its hairless leaflets and leaf base and by the lack of teeth along the margins of its leaflet base.

Purple cinquefoil (*Potentilla palustris*)

• 8 to 20 inches
• mid-season
• wet, swampy habitats

The circumboreal purple cinquefoil is rare in Tahoe, but common in the marshy habitats of Grass Lake and Osgood Swamp. It grows from matted, often floating *stolons*, with ascending flower stalks. The leaves are light green and pinnate to palmate, with two to three toothed leaflets per side. The unusual flowers have dark red-purple petals, stamens, and pistils. The petals are half the size of the star-shaped, green sepals.

Bitter cherry (Prunus emarginata)

• 4 to 12 feet
• early to mid-season
• semi-moist to dry rocky slopes, forest openings

Bitter cherry is a common associate of other dry-adapted montane shrubs up to 8,500 feet. It has silvery bark and small, deciduous leaves, which are elliptic, slightly serrated, and often folded halfway like a canoe. The white flowers bloom in tight, round-topped clusters at the stem tips. The petals are deciduous (like a plum tree), and there are numerous, exserted, yellow-anthered stamens. In contrast to some of its *Prunus* relatives, the bitter cherry fruit is not tasty. The foothill species, **western chokecherry (*Prunus virginiana* var. *demissa*)**, is uncommon on the western lake shore and along the lower Truckee River. Chokecherry has similar leaves, but is distinguished by its hanging inflorescence of over twenty individual flowers. Desert peach (*P. andersonii*), a species formerly found at low elevations around the south shore, is now considered to be extirpated in Tahoe.

Graceful cinquefoil

Purple cinquefoil

Bitter cherry

Bitterbrush

Bitterbrush (*Purshia tridentata*)

- 1 to 3 feet
- early season
- dry forest openings, slopes

Bitterbrush is common up to 9,500 feet, most often on volcanic soils. It becomes ever more abundant as one moves east through Jeffrey pine forest over the Carson Range into the sagebrush scrub community of the Great Basin. It has distinctive small, greenish, three-lobed leaves and many yellow flowers that bloom early, dropping their somewhat disorganized petals by mid-summer. Also known as antelope brush, bitterbrush is a favorite forage of mule deer.

Mountain rose (*Rosa woodsii* var. *ultramontana*)

- **2 to 7 feet**
- **mid-season**
- **semi-moist forest openings**

The genus *Rosa* comprises over 100 species worldwide, with many multi-petaled horticultural derivatives. This widespread shrub is occasional, often locally abundant, in semi-moist forests in the upper montane zone. The flowers have five pink petals surrounding numerous pistils and stamens. The leaves are odd-pinnate with toothed, elliptic leaflets. The branching, often thicket-forming stems have small thorns. The fruit consists of achenes surrounded by a fleshy, reddish hypanthium known as a rose hip. Although not as sweet as the fruits of related species in *Prunus* and *Rubus*, rose hips are considered the greatest natural source of vitamin C.

Thimbleberry (*Rubus parviflorus*)

- **1 to 2 feet**
- **mid-season**
- **semi-moist, shady forest**

Thimbleberry is abundant on northern and eastern exposures in the upper montane zone, often forming low monocultural understories broken only by the interspersed trunks of large conifers. The large, palmate leaves have five toothed lobes. The one- to three-inch wide, white flowers have numerous pale yellow pistils and stamens. The sweet fruits turn bright red by late season and can be eaten off the stem. *Rubus* includes the native California blackberry (*R. ursinus*), salmonberry (*R. spectabilis*), and raspberries (*R. glaucifolius* and *R. leucodermis*), none of which occurs in the Basin.

Sibbaldia (*Sibbaldia procumbens*)

- **1 to 4 inches**
- **mid-season**
- **semi-moist meadow edges, rocky slopes**

This mat-forming, high-elevation plant is occasional in the subalpine zone. The ternate leaves have wedge-shaped leaflets that are generally three-toothed at the tip. The small, star-shaped flowers bloom low to the ground with tiny, oblanceolate, yellow petals and larger, triangular, green sepals and linear

Mountain rose

Thimbleberry

Sibbaldia

Mountain ash

bractlets. The genus is widespread in mountainous areas of northern Europe and Asia.

Mountain ash (*Sorbus californica*)

- 5 to 12 feet
- early to mid-season
- stream edges, moist slopes

Mountain ash is a common component of the mid- to higher-elevation riparian community. The leaves are odd pinnate with seven to nine ovate, toothed leaflets per side. The many small white flowers grow in a panicle. As the season wears on, the green leaves turn to peach and the flowers become bright red berries, adding colorful highlights to fall hiking trips. This species is unrelated to true ashes (*Fraxinus* in the olive family, Oleaceae).

Mountain spiraea (*Spiraea densiflora*)

• 8 to 36 inches
• mid-season
• semi-moist, rocky habitats

Mountain spiraea is common in a variety of habitats from 7,000 up to 10,000 feet. It has soft, ovate, slightly toothed leaves and dense, flat to round-topped, rose-colored flower clusters that sparkle in the fading summer light. The individual flowers have five tiny, erect sepals and five spreading petals with numerous, strongly exserted stamens. Cultivated species of spiraea are common in commercial horticulture. The name comes from the Greek word for a plant used in garlands.

MADDER FAMILY (RUBIACEAE)

The large madder family has over 10,000 species worldwide, mostly in the tropics. Its members contribute such well known products as coffee, quinine, and the ornamental gardenias. Tahoe's two genera, *Kelloggia* and *Galium*, are characterized by opposite or whorled leaves and small, three- to five-lobed flowers. The ovary is inferior, and the fruit consists of two small, fused nutlets covered with dense hooked hairs, which aid in dispersal.

Sweet-scented bedstraw (*Galium triflorum*)

• 4 to 16 inches
• early to mid-season
• moist, shaded habitats

Galium species are low-growing, creeping herbs with whorled leaves and tiny, inconspicuous flowers. The common name derives from the historical use of dried plants as filler in pillows and mattresses. This common low-elevation perennial has four-lobed flowers and whorls of six ovate leaves with pointed tips. The non-native annual **goose-grass** (*G. aparine*) grows in similar habitats, aided by small hooked hairs along its stem and leaves, which attach to adjacent vegetation. It has small, four-lobed flowers and narrow, oblanceolate leaves in whorls of six to eight. Two native species have three-lobed flowers. **Matted bedstraw** (*G. trifidum*) occurs along streams or in shady meadows. This six- to sixteen-inch perennial has linear to obovate leaves in whorls of four to six and, like goosegrass, relies on tiny hooked hairs to gain support. **Mountain bedstraw** (*G. bifolium*) is under six inches tall with tiny, solitary flowers and leaves in whorls of two to four. This delicate annual has a wide distribution in Tahoe, from low-elevation stream-banks to moist, forested slopes into the subalpine zone.

Mountain spiraea

Gray's bedstraw

Sweet-scented bedstraw

Gray's bedstraw (*Galium grayii*)

• 2 to 9 inches
• early to mid-season
• open, rocky slopes, ledges

The high-elevation Gray's bedstraw is relatively common in the north Basin on east- or north-facing talus slopes above 8,000 feet. More dry-adapted than its lower-elevation relatives, this northern California species has distinctly fleshy, almost succulent, rounded leaves, which grow in whorls of four. The four-lobed flowers bloom quickly following snowmelt. The fruits consist of a pair of fused nutlets with long brown hairs, giving the mature plant a glowing appearance as the sun fades over the ridgetops. The southern Sierra **alpine bed-**

straw (*Galium hypotrichium* ssp. *hypo-trichium*) is less common in similar habitats in the south Basin. It has smaller fruits with yellowish hairs and flowers that are well exserted past the leaves.

Kelloggia (*Kelloggia galioides*)

- **6 to 16 inches**
- **early to mid-season**
- **semi-shaded forest to open rocky slopes**

This common species occurs in a variety of habitats into the subalpine zone. It has opposite leaves and small, four- (sometimes five-) lobed flowers. The nutlet fruits have small, hooked hairs similar to those of *Galium*, from which this plant obtains its species name. It is perhaps best recognized by its four angled branches, resulting in a tangle of thin stems, dark green, lanceolate leaves, and flowers that appear to hang suspended in space. *Kelloggia* is named after Albert Kellogg, a frequent nineteenth-century visitor to Tahoe and founder of the California Academy of Sciences.

WILLOW FAMILY (SALICACEAE)

This family consists of only two genera, *Populus* and *Salix*, with over 300 species occurring in northern temperate climates. The willow family provides the only native, non-conifer tree species occurring in Tahoe (except for a few, large canyon live oaks). Most willow family members are moisture-loving and dioecious, having unisexual flowers on separate male and female plants. The flowers have no petals or sepals, instead forming a narrow, dense, many-flowered inflorescence known as a *catkin*. Flowers in catkins are typically wind-pollinated, but many *Salix* species also rely on insect pollination. Species are often well suited to riparian habitats. Many have the ability to survive periodic flooding by growing adventitious, above-ground roots and transporting oxygen through highly developed tissue on stems and leaves. Willow branches or stems broken off by high flows can self-root downstream once water levels have subsided. The family's reproductive cycle also has adapted to the riparian environment, in which wind currents disperse both pollen and feathered seeds up and down the stream corridor.

Black cottonwood (*Populus balsamifera* ssp. *trichocarpa*)

- **up to 100 feet**
- **early season**
- **river, stream corridors**

Populus, which consists of aspen, cottonwoods, and poplars, is composed of wind-pollinated trees with pendant catkins and elliptic

Kelloggia

Black cottonwood

Quaking aspen

to triangular or ovate leaves. Black cottonwood is Tahoe's largest non-conifer, occurring along rivers and streams below 7,500 feet. It has furrowed, gray bark and narrow to widely ovate leaves with slightly serrated margins and an acute tip. The catkins hang with many individual flowers on cup-like disks. In mid-summer the female catkins go to seed, casting flurries of white cotton puffs into the wind. Native Americans made brown dye from its sap and also ate the inner bark as a supplement when food supplies were low. Cottonwood is a favorite of beavers, often resulting in a haphazard maze of felled trees and diverging stream channels.

Quaking aspen (*Populus tremuloides*)

- up to 50 feet
- early season
- moist meadows, slopes, stream margins

This widespread, circumboreal species is common in moist areas in the upper montane zone, particularly in the east Basin. The leaves are ovate with a rounded base and tapered tip. Each leaf is connected by a flattened petiole, which causes the leaves to quiver in the wind like muted chimes. Aspen has a slender, greenish white trunk, which itself may photosynthesize in late fall and early spring when leaves are absent but temperatures are warm enough for productive metabolic activity. Young aspens are rare. The trees flower infrequently, and seedlings require unusual moisture levels for successful establishment. Instead, aspens typically reproduce asexually from underground, horizontal roots, resulting in large patches of cloned trees growing side by side. As nighttime temperatures dip below freezing in fall, the leaves stop producing chlorophyll, unmasking carotene and xanthophyll pigments that showcase bright autumn colors. Clonal aspen populations may be distinguished from one another in the fall based on leaf color variations among different groups of trees.

Arctic willow (*Salix arctica*)

- 1 to 4 inches
- early season
- moist, grassy seeps

Willows are abundant, moisture-loving shrubs that are difficult to identify because of the wide variation in character traits within individual populations and the frequent hybridization between species. Arctic willow is easy to recognize, based on its prostrate growth form and upright catkins, which rise only inches above the grassy terrain. This southern Sierra species occurs on seepy slopes and along lake and stream margins in the subalpine zone at Carson Pass. Willow stems have been used for centuries in basket making. The inner bark is also a source of salicylic acid, a naturally occurring aspirin.

Eastwood willow (*Salix eastwoodiae*)

- 2 to 10 feet
- early season
- moist seeps, stream banks, meadows

Eastwood willow is locally abundant, from low elevations into the subalpine zone, where it may form dense mats in moist environments. The elliptic leaves typically have finely serrated margins and sparsely interwoven, matted hairs, although older leaves may be hairless. There are often small glands along its petioles and leaf margins. This species is named for Alice Eastwood, curator of botany at the California Academy of Sciences and a frequent visitor to Tahoe in

Arctic willow

Lemmon's willow

Eastwood willow

the early 1900s. **Gray-leaved Sierra willow** (*Salix orestera*) is occasional at higher elevations up to 9,500 feet. This species has distinctive, smooth-margined, non-glandular leaves with dense, silky-gray hairs on both surfaces.

Lemmon's willow (*Salix lemmonii*)

• 4 to 12 feet
• early season
• wet meadows, stream banks

Tahoe's most common species has a shrubby appearance, yellow to reddish brown, slender twigs, and greenish, lanceolate leaves, which may have a *glaucous* coating on the underside. Younger leaves may have small, silky white to rust-colored hairs, while older leaves tend to be hairless. The uncommon **Jepson's willow** (*Salix jepsonii*) has dark-purple stems and elliptic to slightly oblanceolate leaves, which are silky-white-hairy on the underside, even when mature. This species is named after Willis Linn Jepson, principal author of the original *Jepson Manual*, and frequent Tahoe visitor. Occasional at lower elevations, **Geyer's willow** (*S. geyeriana*) has minutely hairy, white-silvery, elliptic leaves and shorter (1 to 2 cm.), sub-spherical inflorescences, noticeably distinct from the longer, narrower catkins of other Tahoe species. The eastern Sierra **narrow-leaved willow**

(*S. exigua*) occurs in the far north and south Basin in moist, low-elevation habitats. This species is easily recognized by its long, linear leaves.

Shining willow (*Salix lucida* ssp. *lasiandra*)

• 6 to 30 feet
• early season
• moist stream margins, meadows

This common species grows as a tree at lower elevations and as a shrub into the subalpine zone. It has large, yellowish catkins and still larger (5 to 17 cm. long), lanceolate leaves, which are hairless and shiny dark green on the upper surface and somewhat *glaucous* beneath. Shining willow may be identified by the small glands at the base of its leaf blades and by its rounded, glandular stipules, which clasp the base of the leaf nodes.

Scouler's willow (*Salix scouleriana*)

• 10 to 30 feet
• early season
• white fir forest, semi-moist slopes

This large shrub to small tree is Tahoe's most drought-adapted willow, often occurring in relatively dry openings in lower-elevation forest, as well as in moist areas along stream banks, lakes, and meadows. Scouler's willow is best

identified by its strongly oblanceolate leaves, which are consistently widest past the leaf midpoint.

SAXIFRAGE FAMILY (SAXIFRAGACEAE)

This family of perennial herbs is characterized by palmately veined leaves and small flowers that bloom on a leafless, usually upright stem. Each flower has five sepals, five petals, five or ten stamens, and two pistils. As the ovaries ripen, they rise up and turn reddish, eventually forming a two- (sometimes four-) chambered fruit capsule. The family is well represented in Tahoe, with many species in six genera. Saxifragaceae has approximately 600 species worldwide, mostly in cool, temperate, mountainous regions.

Sierra bolandra (*Bolandra californica*)

• 6 to 24 inches
• mid-season
• moist, rocky ledges, streams

This rare species occurs in wet, rocky habitats below 8,000 feet in the southwest Basin, from Emerald Bay to Carson Pass. The reniform leaves have three to seven toothed lobes. The flowers grow in a panicle on stems with clasping, leaf-like bracts. The narrow, recurving flower petals are green-

Shining willow

Sierra bolandra

Scouler's willow

Pink alumroot (*Heuchera rubescens*)

• **4 to 20 inches**
• **mid-season**
• **rocky ledges, crevices**

This common species occurs on open, rocky cliff sides into the sub-alpine zone. It has broadly ovate, palmately lobed basal leaves and numerous, small flowers that bloom in a raceme on a long, leafless, erect to leaning stem. The calyx is fused into a white to pink, bell-shaped *hypanthium*, out of which project five tiny whitish petals and five stamens. The ovary is half superior, meaning its lower half is enveloped by the fused flower parts. Tahoe populations are typically referred to as *Heuchera rubescens* var. *glandulosa.* The cen-

ish with purple margins. Unlike the similar alumroot, the ovary is superior and the five stamens are not exserted past the fused calyx.

tral Sierra *H. r.* **var.** *rydbergiana* has been found in Carson Pass. The common foothill species, *H. micrantha* has been found in the Glen Alpine area in moist, semi-shady, rocky habitats. This species is distinguished by its tiny, *radial* flowers, with equal-sized calyx lobes. Alumroot was named for its roots, which were historically used by Native Americans and settlers as a soothant for external injuries.

Rock star (*Lithophragma glabrum*)

• **3 to 10 inches**
• **early season**
• **semi-moist, rocky habitats**

This attractive plant blooms early with the buttercups in forest openings, along meadow edges, and on open, rocky slopes up to 9,000 feet. The basal and smaller stem leaves are palmately compound, the three leaflets lobed and toothed. The distinctive white to pink flower petals are deeply three- to five-lobed in an irregular fashion.

Brewer's bishop's cap (*Mitella breweri*)

• **4 to 12 inches**
• **early to mid-season**
• **moist, semi-shady stream banks, lake and marsh edges**

This common species occurs in moist, semi-shady habitats up to 9,000 feet. It is named for its fruit capsule, which resembles the tall,

pointed hat with two peaks worn by a bishop. It has rounded, slightly reniform, toothed basal leaves and an inflorescence that blooms in a raceme from bottom to top on an erect to ascending stem. The unmistakable flowers have tiny, yellow-green petals, each with five to nine delicate, linear lobes per side. The petals radiate out from a centrally fused, saucer-shaped calyx. The resulting floral structure has given rise to the alternative common name of snowflake for various *Mitella* species. Like many plants in Tahoe, this species was named after William H. Brewer, state botanist on the first California Geologic Survey. The rare **five-point bishop's cap** (*Mitella pentandra*) has been found in shady, moist habitats in the west Basin, at Sagehen Creek, and near Carson Pass. It is distinguished by its petals, which are opposite, instead of alternate, to the five stamens.

Fringed grass of Parnassus (*Parnassia fimbriata*)

• **6 to 20 inches**
• **late season**
• **rocky stream banks**

This uncommon species occurs in moist, oxygenated habitat in the northwest Basin. The large flowers have ovate, fringed petals, five stamens, and five infertile stamens called *staminodes*. The round, clasping basal leaves have reniform

Pink alumroot

Rock star

Brewer's bishop's cap

Fringed grass of Parnassus

bases. The more widespread **grass of Parnassus** (*Parnassia californica*) occurs into the subalpine zone along streams and rocky meadow margins. It has similarly large flowers, with unfringed petals, and ovate to lanceolate basal leaves with tapered bases. Parnassus flowers are good examples of floral evolution. The original ten stamens were *reduced* to five, and the infertile staminodes developed minutely fringed margins that entangle nectar-seeking insects long enough for pollen to be exchanged.

Bud saxifrage (*Saxifraga bryophora*)

• **2 to 10 inches**
• **mid-season**
• **semi-moist, sandy habitats**

The inconspicuous bud saxifrage is common in Tahoe from 7,000 feet throughout the subalpine zone. The slightly toothed basal leaves are linear to elliptic. The small flowerbuds nod downward on hanging pedicels from an erect to ascending stalk, blooming from top to bottom into considerably larger, showy, white flowers that have two yellow spots near the base of each petal, ten stamens, and two ovary chambers. As is true of most species of *Saxifraga*, the flowers are protandrous, avoiding self-fertilization by delaying the development of female parts. Once the stamens have dispersed their pollen, the two-beaked ovary begins to rise, offering the now receptive stigmas to pollinators from other plants.

Brook saxifrage (*Saxifraga odontoloma*)

• **8 to 20 inches**
• **mid-season**
• **shady, rocky stream margins**

This characteristic plant of narrow, rocky stream courses is relatively common from low elevations up to 9,000 feet. The distinctive, large basal leaves are round and reniform with sharply toothed margins. The inflorescence blooms in a cyme on stems that lean out precariously over the open, rushing water. Each delicate flower has reddish, reflexed sepals, yellow-dotted, white petals, and spreading, clubshaped stamens with red anthers.

Bog saxifrage (*Saxifraga oregana*)

• **10 to 40 inches**
• **early to mid-season**
• **wet to moist meadows, boggy seeps**

Bog saxifrage is common throughout the upper montane zone, occurring abundantly at Sagehen Creek and Tahoe Meadows. It has large, oblanceolate, sometimes slightly toothed basal leaves and tiny flowers, which cluster together on short peduncles from a thick, erect, central stalk. Two similar species, **peak saxifrage** (*Saxifraga nidifica* var. *nidifica*) and **Sierra saxifrage** (*S. aprica*), prefer higher-elevation, well draining, moist habitats, including rocky slopes,

Bud saxifrage

Brook saxifrage

Bog saxifrage

slightly toothed. The flowers bloom in small clusters atop a thin, six- to eighteen-inch stem. The reflexed to spreading sepals are light green. The elliptic to roundish petals are one to two millimeters long. Sierra saxifrage has obovate to elliptic, sometimes reddish basal leaves, a six-inch stem, and somewhat reddish sepals that tend to be erect. The petals are ovate and slightly larger.

Alpine saxifrage (*Saxifraga tolmiei*)

- **1 to 6 inches**
- **mid-season**
- **semi-moist, rocky to sandy slopes**

Common in the Pacific Northwest, alpine saxifrage is occasional in the

ledges, and subalpine stream beds. Peak saxifrage has small, ovate to triangular basal leaves that are

southwest Basin, on snowmelt-fed, rocky ledges and fell fields from 8,500 to 10,500 feet. It has small, fleshy, elliptic leaves, which cluster on trailing stems from a woody base. The many flowers bloom in small heads off erect, occasionally branching stalks. The white petals are linear to oblanceolate.

FIGWORT FAMILY (SCROPHULARIACEAE)

The figwort family is represented in Tahoe by eleven native genera and numerous species, consisting of mostly perennial herbs with showy, irregular flowers. The five fused sepals form a lobed calyx. The five petals are fused into an upper lip with two lobes and a lower lip with three lobes. Flowers have a superior ovary and typically contain four stamens, in two pairs, with a fifth infertile stamen (staminode) sometimes present. Leaves are generally alternate and entire. The variety of floral structures in the figwort family is attributable to the co-evolution of its members with pollinators such as bees and hummingbirds. The family's diversity, which spans such disparate genera as *Limosella*, *Castilleja*, *Mimulus*, *Penstemon*, *Pedicularis*, *Veronica*, and *Ortho-carpus*, suggests that Scrophulariaceae may be split into several families in the near future. The family is named after the disease scrofula, a tubercular condition of the lymphatic glands, which some figwort species' roots were thought to resemble and, consistent with seventeenth-century medical knowledge, also used to cure.

Applegate's paintbrush (*Castilleja applegatei*)

- 6 to 18 inches
- early to mid-season
- open slopes, forests

Castilleja species are parasites that tap the root systems of neighboring plants to obtain water and nutrients. This plant is abundant into the alpine zone. Like all species of *Castilleja*, it has small, inconspicuous flowers surrounded by colorful bracts that assume the role of attracting pollinators. The green leaves are distinctively wavy-margined. Normally red-bracted, bracts of some Carson Pass populations may range from pale yellow to whitish pink. The rare **frosty paintbrush** (*C. pruinosa*) occurs on dry, forested slopes up to 8,000 feet. It has an orange-red inflorescence, straight-margined leaves, and dense stem and leaf hairs, which are distinctly branched. The hairs reduce moisture loss and, as is true of many alpine plants, give the plant a frosty gray appearance. **Linear-leaved paintbrush** (*C. linariifolia*) occurs along subalpine ridges in the southeastern Carson

Alpine saxifrage

Giant red paintbrush

Applegate's paintbrush

Range. It has narrow leaves with rolled margins and large, somewhat exserted flowers. The calyx has deeper lobes in the front than in the back of the flower. This common western species is the state flower of Wyoming.

Giant red paintbrush (*Castilleja miniata* ssp. *miniata*)

- **16 to 32 inches**
- **mid-season**
- **moist to semi-moist habitats**

This tall paintbrush is abundant up to 9,000 feet in moist, well drained soils. It has pointed, red to yellow bracts, and entire (unlobed), lanceolate leaves. As is true of all red-bracted paintbrushes (and of scarlet gilia and California fuchsia), patches of this plant in bloom are a good place to watch for hummingbirds.

Small-flowered paintbrush (*Castilleja parviflora*)

- **8 to 16 inches**
- **mid-season**
- **moist to dry meadows**

This attractive species is abundant in subalpine meadows of the Desolation Wilderness, and occasional elsewhere in the southwest Basin.

The flower head bracts are deep red, with a strongly two-lobed stigma. More common in the central and southern Sierra, **Lemmon's paintbrush** (*Castilleja lemmonii*) is rare in Tahoe, occurring in moist subalpine meadows near Castle Peak, Tahoe Meadows, and in the Desolation Wilderness area. It is distinctive for its brilliant magenta bracts and low, two- to six-inch stature.

Alpine paintbrush (*Castilleja nana*)

• **2 to 6 inches**
• **mid-season**
• **dry, rocky slopes, summits**

This variably colored species is common above 7,500 feet. The bracts range from green to grayish purple to rust red. Alpine paintbrush shares its high-elevation habitat with hairy paintbrush, from which it is distinguished by its pointed, green-margined bract lobes and loose-fitting seed coats, which shake like a rattle later in the growing season.

Hairy paintbrush (*Castilleja pilosa*)

• **1 to 12 inches**
• **mid-season**
• **drying meadows, rocky alpine summits**

This species grows in matted form on the highest summits in the Basin, but also occurs in drying meadows at lower elevations, where it may reach a foot in height. Like alpine paintbrush, it prefers sagebrush (*Artemisia spp.*) as a host for its root parasitism. Hairy paintbrush is distinctive for its bract lobes, which are generally rounded and distinctly white-margined. In addition, its seed coats fit tightly and do not rattle when shaken.

Slender paintbrush (*Castilleja tenuis*)

• **2 to 8 inches**
• **mid-season**
• **moist to dry meadows, slopes**

Formerly known as hairy owl's clover, this diminutive annual has small yellow flowers protruding from greenish bracts along a short, erect stem. The flowers are pollinated by bees, as are those of other paintbrushes with non-red bracts. Slender paintbrush is especially common in the north Basin, often growing in large populations where there has been sufficient moisture.

Blue-eyed mary (*Collinsia torreyi*)

• **2 to 4 inches**
• **early season**
• **open slopes, forests**

This common name refers to several annual species with showy, two-lipped flowers of blue and white (or sometimes lavender). In years of favorable moisture, blue-eyed marys may carpet whole ar-

Small-flowered paintbrush

Hairy paintbrush

Alpine paintbrush

Slender paintbrush

eas with a vast twinkling of color far out of proportion to the plants' tiny stature. Common on open, sandy slopes, this species has dense glandular hairs, a common adaptation to dry habitats. It survives into the subalpine zone in Tahoe's harsh environment by exploiting plentiful moisture early in the growing season, flowering and going to seed before the near-surface soil dries completely. The smaller-flowered *Collinsia parviflora* oc-curs in moister forest or meadow habitats. It has few or no glandular hairs.

Bird's beak (*Cordylanthus tenuis*)

• 1 to 3 feet
• mid- to late season
• open forest and slopes

This annual root parasite occurs in a variety of habitats from open, rocky terrain to semi-shaded forest. It is named for the manner in which the upper and lower lips of its corolla fold together. The solitary, white and yellow flowers have a purple splotch on the upper lip and occur in groups of four to seven on multi-branched stems. This earth-toned, wispy plant blends well with its environment, perhaps accounting for its infrequent sighting in Tahoe. It is common along hiking trails in Glen Alpine and Shirley Canyon.

Lemmon's keckiella (*Keckiella lemmonii*)

• 2 to 4 feet
• mid- to late season
• dry, open slopes

This subshrub is locally common below 8,000 feet on south-facing, forested or chaparral hillsides in the northwest Basin. The small leaves are oval-shaped and grow opposite along the long, ascending stems, from which bloom many small, delicately colored flowers. The stamens are densely hairy at the base, a character that distinguishes *Keckiella* from *Penstemon*, the genus to which it formerly belonged. *K. breviflora* makes a rare Tahoe appearance on steep, dry slopes near Glenbrook. This similar-sized shrub has white to pinkish flowers and oblanceolate leaves with serrate margins.

Mudwort (*Limosella acaulis*)

• 1 to 2 inches
• mid-season
• muddy shorelines

This matted annual is uncommon, but at times locally abundant, on the edges of lakes and swamps at lower elevations. The half-inch to two-inch leaves are long and linear, widening slightly at the end into a spoon shape. Mudwort has tiny white flowers with five, equal-sized, rounded lobes. The four stamens are slightly exserted. The genus name means "of muddy places."

Brewer's monkeyflower (*Mimulus breweri*)

• 1 to 6 inches
• early season
• moist open habitats

One of Tahoe's smallest *Mimulus* species, Brewer's monkeyflower is occasional up to 9,000 feet in seep areas where shallow soil depth does not permit perennial competitors to monopolize resources. This annual species has tiny pink flowers that sit atop relatively long (2 to 15 mm.) flower stems (*pedicels*). The similar-sized **Sierra monkeyflower**

PLANT DESCRIPTIONS 269

Blue-eyed mary

Bird's beak

Lemmon's keckiella

Mudwort

(*M. leptaleus*) has small pedicels, typically shorter than the reddish, glandular calyx. The pink flowers have a lighter-colored throat with dark spots. *Mimulus* is Latin for little mimic. The genus was named after the shape of the corolla, which was thought to resemble the face of a monkey.

Common monkeyflower (*Mimulus guttatus*)

• **1 to 30 inches**
• **early to mid-season**
• **wet to moist habitats**

California's most widespread monkeyflower, this species occurs in a variety of moist habitats, especially seeps, below 8,000 feet. The more typical, perennial growth form has large, irregular, yellow flowers that bloom in groups of five or more on a raceme punctuated by pairs of rounded, slightly toothed, leaf-like bracts. (The leaves were commonly eaten in salads by Native Americans and early settlers.) The bracts clasp the stem at the base of each flower pedicel. As the fruits develop, the calyx swells asymmetrically, with the lower lip noticeably larger than the upper. The annual form grows under four inches, with one to three small flowers punctuated by a red spot on the lower corolla lip. The tiny, yellow-flowered annual, *Mimulus suksdorfii*, found at Sagehen Creek and at Carson Pass, is distinguished by its linear to ovate stem leaves and a calyx that does not swell in fruit.

Lewis' monkeyflower (*Mimulus lewisii*)

• **1 to 3 feet**
• **mid- to late season**
• **moist seeps, streams**

Lewis' monkeyflower is common, especially on the west side, up to 9,000 feet. It is easily recognized by its large stature and lavender to pink flowers with clear yellow lines on the lower lip of the corolla, which guide pollinators toward the nectar deep inside the flower tube. The large, elliptic leaves are palmately three- to five-veined.

Dwarf monkeyflower (*Mimulus mephiticus*)

• **1 to 6 inches**
• **early to mid-season**
• **dry, sandy slopes and flats**

This short-statured annual is occasional in dry, sandy habitats into the subalpine zone. It may be distinguished from similar species by its exserted (past the corolla lobes) stigma and its densely hairy, glandular-sticky leaves and calyx. Flowers are typically magenta, especially in the north, but sometimes yellow toward the south, with both forms occasionally occurring together. (Yellow-flowered forms are Tahoe's only yellow monkeyflower with pedicels shorter than the flower tube.) **Musk monkeyflower** (*Mimulus moschatus*) grows in a variety of moist to drying habitats, often in partial shade. This yellow-

Brewer's monkeyflower

Common monkeyflower

Lewis' monkeyflower

Dwarf monkeyflower

flowered perennial has corollas with equal-sized lobes, lending flower the appearance of regular-ity. Oblong to ovate stem leaves, though variable, all have a slimy texture when rubbed.

Primrose monkeyflower
(*Mimulus primuloides*)

• 3 to 6 inches
• mid-season
• wet to moist meadows, seeps

This diminutive perennial occurs in wet to moist habitats up to 9,500 feet. The slightly irregular flowers are shallowly lobed with small red spots along the lower lip base. The flowers bloom individually atop short, erect, leafless stems (like a primrose) that rise from a compact rosette of small, densely hairy basal leaves. The widespread **downy monkeyflower** (*Mimulus pilosus*) is rare on low-elevation, moist, sandy slopes along the southeastern lake shore. This delicate, one-to twelve-inch-tall annual is covered by long, soft hairs. The minuscule yellow flowers have two maroon spots on the lower corolla lip.

Mountain monkeyflower
(*Mimulus tilingii*)

• 2 to 10 inches
• mid-season
• moist seeps, stream margins

Mountain monkeyflower is common in rocky seep areas from 7,000 to 10,000 feet. It differs from the similar *Mimulus guttatus* in its bractless inflorescence, which usually terminates in one to three flowers, and its general preference for higher elevations.

Torrey's monkeyflower
(*Mimulus torreyi*)

• 2 to 8 inches
• early to mid-season
• semi-moist to dry forest openings, slopes

This minutely hairy annual is more common in the north Basin, where it blooms at low elevations, often forming huge mats of brilliant magenta. The lower lip of each flower has two prominent, gold nectar stripes. It differs from *Mimulus mephiticus* in its generally non-glandular herbage and its stigma, which does not protrude past the corolla lobes.

Copeland's owl's clover
(*Orthocarpus cuspidatus* ssp. *cryptanthus*)

• 4 to 16 inches
• mid-season
• open volcanic scrub

Owl's clovers are annual root parasites distinguishable from paintbrushes (*Castilleja*) by their generally less colorful bracts and upper and lower corolla lips that completely enclose the inner flower parts. This species is locally common on volcanic slopes in the north and south Basin, typically in association with mountain sagebrush. The upper bracts are purplish pink. The flower has a pinkish upper lip and whitish lower lip. The genus name is Greek for upright fruit.

Primrose monkeyflower

Torrey's monkeyflower

Mountain monkeyflower

Copeland's owl's clover

Little elephant's head
(*Pedicularis attollens*)

- 4 to 12 inches
- mid- to late season
- moist meadows, grassy slopes, stream margins

Pedicularis species are root parasites, best characterized by alternate, deeply divided leaves. Elephant's heads have pink to purple flowers that resemble the ears and trunk of an elephant. This species has small, brilliant pink flowers in which the "trunk" lobe spirals back into the corolla. It can be locally abundant in moist subalpine habitats, blooming slightly later in the season than its larger cousin.

Elephant's head (*Pedicularis groenlandica*)

- 8 to 16 inches
- mid-season
- wet, moist meadows, seeps

This moisture-loving species is common up to 8,500 feet. It differs from *Pedicularis attollens* in its darker purple flowers and long, extended "trunk" petal lobe. Elephant's head is buzz-pollinated by bumblebees, which coax pollen from narrow anther openings near the base of the trunk petal by vibrating their wings. At the next flower, the bee rubs its pollen-covered underside against the receptive stigma, which extends out of the end of the trunk petal lobe.

Pinewoods lousewort (*Pedicularis semibarbata*)

- 2 to 6 inches
- early season
- dry, shady to open forest, slopes

Pinewoods lousewort is common throughout Tahoe's upper montane forest communities. This prostrate species has inconspicuous, yellow, tubular flowers with reddish or purplish tips. The flowers are borne amidst dark green, basal leaves. The genus and common names are derived from the traditional belief that livestock that consumed European lousewort were cured of lice.

Azure penstemon (*Penstemon azureus* var. *azureus*)

- 8 to 20 inches
- early to mid-season
- rocky slopes, forest openings, roadsides

Penstemons are named for their five stamens, the fifth of which, the *staminode*, is sterile. (The staminode is often hairy, thus giving rise to another common name, beard tongue.) Penstemons have opposite leaves and tubular flowers that vary according to the size of the bee pollinators that tend to visit each species. A hand lens can aid in keying out the species based on clearly distinguishable characteristics of the floral parts. Azure penstemon occurs in the northwest Basin from Donner Summit to upper Pole Creek. This striking, large-flowered species has short, heart-shaped, clasping leaves and deep yellow buds, which give the blooming inflorescence a collective blue and gold appearance. The flowers have neither hairs nor glands. The anthers open with pollen at the point of connection to the filament, leaving the end portions of the anthers closed.

Little elephant's head

Elephant's head

Pinewoods lousewort

Davidson's penstemon (*Penstemon davidsonii* var. *davidsonii*)

- 1 to 6 inches
- mid-season
- sandy to rocky slopes, summits

Azure penstemon

On or near the summits of Tahoe's highest peaks, one finds this rare alpine penstemon, a matted plant with large, blue-violet flowers. The creeping woody stems have small, upwardly reduced, rounded leaves.

Davidson's penstemon differs from showy penstemon in its pale yellow, hairy staminode and densely woolly anthers, which shed pollen the full length of their margins. This species is thought to be an alpine derivative of mountain pride (*Penstemon newberryi*), having evolved into a separate species as a result of reproductive isolation during periods of glaciation.

Hot rock penstemon (*Penstemon deustus*)

- **6 to 16 inches**
- **early to mid-season**
- **sandy to rocky slopes, ledges, forest openings**

Hot rock penstemon is common on rock outcroppings and amidst Jeffrey pine forest from low elevations up to 8,500 feet. This species has creamy white flowers and slightly toothed, small, leathery leaves. Fine, dark-lined veins on the lower petals give the plant its species name, which means burnt. The tall (up to three feet), red-flowered **Eaton's penstemon** (*Penstemon eatonii*) occurs in the south Basin along Highway 89 near Meyers, perhaps as a result of CalTrans reseeding projects. This species is normally found in desert mountains to the southeast.

Slender penstemon (*Penstemon gracilentus*)

- **8 to 20 inches**
- **mid-season**
- **dry, semi-shady to open forest**

Occasional into the subalpine zone, slender penstemon is locally abundant amidst the Jeffrey pine forest understories characteristic of the eastern Basin. Also known as graceful penstemon for its delicate stems, leaves, and flowers, this species has a yellow, hairy staminode and glandular flowers and stems. Its whorled inflorescence differs from *Penstemon heterodoxus* by short stems (peduncles) from which groups of flowers bloom. The purple flowers may range into pale lavender in some populations.

Mountain pride (*Penstemon newberryi* var. *newberryi*)

- **6 to 12 inches**
- **mid-season**
- **rocky slopes, open forest**

A matted, woody subshrub, mountain pride is common in a variety of habitats into the subalpine zone. It has glandular, pink flowers, which light up the surrounding granitic terrain. Small, mostly basal leaves are obovate with minutely serrated margins. Like the closely related *Penstemon davidsonii*, its anthers are densely woolly and open along their entire margin.

Davidson's penstemon

Slender penstemon

Hot rock penstemon

Gay penstemon (*Penstemon roezlii*)

• 1 to 2 feet
• mid-season
• dry, rocky slopes, open forest, montane chaparral

This cheerful plant is common in dry habitats in the north and south

Mountain pride

Basin, mostly on granitic outcroppings. Populations in the north tend to have blue flowers and ascending stems. Those in the south have more rose-purple flowers and sprawling stems. It has a hairless staminode and glandular inflorescence. The leaves are linear to lanceolate. The stamens release pollen similarly to azure penstemon, from the point of connection with the filament. This species was formerly considered a subspecies of *Penstemon laetus.* It differs from varieties of that species in its shorter (14 to 22 mm.) corolla.

Meadow penstemon (*Penstemon rydbergii* var. *oreocharis*)

• **8 to 20 inches**
• **mid-season**
• **moist meadows**

Tahoe's only moisture-loving penstemon has a whorled inflorescence of generally hairless, glandless flowers. Meadow penstemon is often the most abundant member of the upper montane, drying meadow community, blooming just prior to bog mallow after the camas lilies have gone to seed. The similar **whorled penstemon** (*Penstemon heterodoxus*) occurs on rocky, snowmelt-fed, subalpine mountainsides. It has distinctly glandular flowers that bloom in a similar whorled inflorescence on a six- to twelve-inch, erect to ascending stem. More common to the east, *P. procerus* var. *formosa* has been found on the summit of Stevens Peak and in the Mount

Rose-Slide Mountain area. It has a semi-whorled, glandless inflorescence of small, blue-purple flowers. The lanceolate to narrowly ovate stem leaves are folded lengthwise along a one- to six-inch stem.

Showy penstemon (*Penstemon speciosus*)

• **2 to 30 inches**
• **mid-season**
• **open, rocky slopes, summits**

This highly variable penstemon is common from lake level to the alpine zone. The flowers are larger (2.5 to 3.5 cm. long), and the leaves thicker and more leathery (often with a waxy blue surface) than those of other Tahoe species. Showy penstemon becomes shorter and more densely flowered with increasing elevation, appearing in matted form on Tahoe's highest mountain summits. It is the only Tahoe species whose anthers open on the ends, away from the point of contact with the stamen filament. The staminode is mostly hairless with some occasional hairs at the tip.

Figwort (*Scrophularia desertorum*)

• **2 to 4 feet**
• **mid-season**
• **semi-open, protected slopes, boulder crevices**

Also known as beeplant, this straggly species is uncommon in par-

Gay penstemon

Showy penstemon

Meadow penstemon

Figwort

tially shaded, protected habitats ranging from open, low-elevation forest to high-elevation boulder fields. Easy to overlook, figwort has large, toothed, triangular leaves and tiny, two-lipped flowers. The short flower tube is a snug fit for preferred bee pollinators.

Woolly mullein (*Verbascum thapsus*)

• 1 to 6 feet
• mid- to late season
• open disturbed areas

This Eurasian native is common in a variety of low-elevation habitats. It has an unbranched inflorescence, erect, woolly herbage, and yellow flowers. Like green gentian, woolly mullein is a biennial, putting forth a basal rosette of leaves in the first year and growing an inflorescence the second. The flowers bloom for a single day and, unusual for a vertical inflorescence, in no particular sequence. The name "mullein" comes from the Latin word for soft. Native Americans made tea from the leaves for sore throats. Early settlers used the leaves as lamp wicks and the stalks as torches.

American speedwell (*Veronica americana*)

• **2 to 12 inches**
• **early to mid-season**
• **wet habitats**

Veronica is a surprisingly large genus in Tahoe, with five species that occur in a variety of moist habitats. Veronicas have a saucer-shaped flower, the upper lip of which is fused into a single lobe, resulting in a four-lobed corolla, unusual for plants in this family. There are two exserted stamens and, in some species, a long exserted style, both characters distinct from other four-petaled flowers. The fruit is a flattened, heart-shaped, two-chambered capsule, often notched at the top. This occasional species can be abundant in stream or semi-shady wet environments at low elevations. It has

a creeping, prostrate stem, which puts forth an inflorescence of up to twenty-five dark blue-violet flowers. The name "speedwell" comes from the tradition in which veronicas were said to have greeted earlier travelers.

Cusick's speedwell (*Veronica cusickii*)

• **4 to 8 inches**
• **mid-season**
• **moist, grassy slopes, seeps**

The rare Cusick's speedwell occurs in moist, grassy subalpine habitats at Carson Pass and in upper Shirley Canyon. It has large flowers and a long (6 to 9 mm.) style. The elliptic to ovate leaves are generally hairless. The annual **purslane speedwell** (*Veronica peregrinus* ssp. *xalapensis*) occurs on muddy lake shores and in meadows into the subalpine zone. In Tahoe, it typically grows under four inches tall with tiny white flowers and linear to spoon-shaped leaves.

Thyme-leaved speedwell (*Veronica serpyllifolia* ssp. *humifusa*)

• **2 to 12 inches**
• **early to mid-season**
• **moist meadow, seeps, streambanks**

This bright-flowered species is common to 9,500 feet. The light blue flowers have dark nectar lines, and the ascending stems have

Woolly mullein

Cusick's speedwell

American speedwell

Thyme-leaved speedwell

sessile, ovate leaves. The heart-shaped fruits are wider than long, with a distinct notch at the top, and barely exceed the persistent calyx. The similar **alpine speedwell** (*Veronica wormskjoldii*) grows erect up to twelve inches in similar habitats above 7,000 feet. The slightly smaller flowers are dark blue with a shorter style. The elliptic to lanceolate leaves are long-hairy. The oval-shaped fruits are longer than wide and greatly exceed the calyx. Both species occur along creeks and in meadows below Mount Rose.

NIGHTSHADE FAMILY (SOLANACEAE)

This large and diverse, mostly South American family has alternate, entire to lobed leaves and flowers with five calyx and corolla lobes, five stamens, a superior, two-chambered ovary, and one style. The fruit is a capsule or berry. The family includes such important economic crops as potatoes, eggplants, peppers, tomatoes, tobacco, and petunias. Many members contain strong, sometimes poisonous, alkaloids, including nicotine (*Nicotiana*) and stramonium (*Datura*). Aware of these chemical properties, Europeans initially refused to accept the tomato as a viable crop plant from the New World.

Dwarf chamaesaracha (*Chamaesaracha nana*)

- 1 to 6 inches
- early to mid-season
- semi-moist, sandy, rocky flats

This uncommon perennial occurs below 9,000 feet, mostly in the north and south Basin. It grows from several sprawling stems with flowers that bloom singly or in axillary clusters of from two to five. The white petals are fused into an open, saucer shape with a greenish yellow center and five free stamens whose filaments are larger than the yellow anthers. The tomato-like leaves are entire to pinnately lobed with wavy margins. The fruit is a spherical berry. Coyote tobacco (*Nicotiana attenuata*), common in the eastern Sierra, was collected years ago at low elevations in the south, but is now considered extinct in Tahoe.

Purple nightshade (*Solanum xanti*)

- 1 to 3 feet
- mid-season
- dry, sandy slopes, forest openings

This occasional, upper montane species has large dark blue to lavender flowers, which hang in small umbel clusters off multi-branched stems. The five stamens have large cylindrical anthers, which grow together in a single, erect, yellowish beak, through which projects the club-like stigma. Flowers produce no nectar; pollination occurs through buzz pollination. Leaves are lanceolate to ovate with occasional lobes at the subcordate base. Fruits resemble small, green tomatoes, each attached to a persistent five-lobed calyx. *Solanum* has an estimated 1,500 species worldwide, including the cultivated potato, *S.*

Dwarf chamaesaracha

Purple nightshade

tuberosum. Purple nightshade is one of several natives of nineteen species occurring in California. The generic name comes from the Latin *solan-um,* which means making quiet, a reference to the sedative properties of some nightshade species.

Stinging nettle

NETTLE FAMILY (URTICACEAE)

This widespread family ranges from annuals to trees with over 700 species worldwide. The small, greenish, wind-pollinated flowers are unisexual with no petals and occur in axillary clusters. The ovary is superior. There is one style and one stigma.

Stinging nettle (*Urtica dioica* ssp. *holosericea*)

• **3 to 7 feet**
• **mid-season**
• **moist stream, meadow edges**

Nettle is occasional (often locally abundant) in moist habitats below 7,500 feet. It grows on an erect, gray-green stem with opposite, lanceolate, toothed leaves. Flowers are unisexual. Male flowers have four equal-sized sepals, while the two outer sepals of female flowers are smaller than the two inner ones. Like many nettle species, the plant is covered with finely pointed hairs, which easily pierce open skin to inject formic acid, causing an intense and lasting stinging sensation. The perceived medical benefits of these stinging hairs gave rise, at one time, to the flogging of patients with nettle branches, an inauspicious beginning for physical therapy treatment. Native Americans boiled nettle leaves and stems as potherbs and used the stem fibers to make cloth, fishing line, and rope.

VALERIAN FAMILY (VALERIANACEAE)

The valerian family has around 300 species worldwide, generally in northern temperate areas and in the South American Andes. It is composed of annual and perennial herbs with whorled basal and opposite stem leaves that are often lobed to compound. The fused flower parts are five-lobed, and the ovary is inferior. There are four genera in California.

California valerian (*Valeriana californica*)

- **10 to 20 inches**
- **early to mid-season**
- **moist to dry habitats**

This abundant species occurs in a variety of habitats up to 9,000 feet, often in volcanic soils. It grows on several erect to ascending stems with deeply lobed leaves and creamy white flowers that bloom from pinkish buds in terminal cyme clusters. The flowers are funnel-shaped and characteristically swollen near the base of the corolla. Each flower has acute to rounded, equal-sized lobes and three exserted stamens. Valerians have a strong, somewhat disagreeable odor; the plants were nevertheless utilized as a source of perfume in sixteenth-century Europe.

VIOLET FAMILY (VIOLACEAE)

The violet family has 600 species worldwide, ranging from annual herbs to shrubs and vines. Violet flowers have five sepals, five petals, five stamens, a superior ovary, and one style. The petals are free and unequal, the lowest usually having a spur or pouch that contains the nectar sought by bee pollinators. The fruit is a three-valved capsule. *Viola* is the only native genus in California. Many violets have some *cleistogamous* flowers, meaning

California valerian

Western dog violet

that they reproduce through self-pollination, without opening. Cleistogamy is an adaptive response to the tendency of violets to bloom early in the season when insect pollinators are not always present. Violets have been prized since antiquity for their delicate colors and fragrances.

Western dog violet (*Viola adunca*)

• **2 to 6 inches**
• **early season**
• **moist, semi-shady forest, stream margins**

Common in the western states, this species is occasional at low elevations throughout the Basin. It has round to ovate basal and stem leaves and pale to deep purple flowers with a spurred lower petal. (It is also known as long-spurred violet.) Flowers are pollinated mostly by bumblebees that follow distinct nectar lines toward the sweet rewards at the base of the petal spur. The eastern Sierra **LeConte's violet** (*Viola sororia* ssp. *affinis*) is rare in wet, grassy seeps at Sagehen Creek and Donner Lake. It has ovate to reniform, toothed basal leaves, no stem, and purple flowers with a much smaller spur.

Beckwith's violet (*Viola beckwithii*)

• **2 to 5 inches**
• **early season**
• **moist, rocky meadows, slopes**

This Great Basin native is locally common in the far north Basin, typically in the transition zone between drying meadows and sagebrush scrub. It also occurs on the southwestern slopes of Red Lake Peak. The showy flowers have light pink to white lower petals, with red-violet veins and yellow bases, and deep maroon upper petals. The distinctive, light green leaves are once-ternate with dissected, linear to lanceolate leaflets.

Stream violet (*Viola glabella*)

• 4 to 12 inches
• early season
• moist stream edges, shady meadows

Common up to 7,500 feet, this species has relatively large, ovate to reniform leaves that occur directly beneath the deep yellow flowers. Each flower has delicately purple-veined lower petals. **Baker's violet** (*Viola bakeri*) occurs in drier, forested and grassy habitats at low to medium elevations. (It is common at Sagehen Creek and Alpine Meadows.) This yellow-flowered species grows up to one foot tall with entire edged (not wavy or jagged), lanceolate leaves.

Maclosky's violet (*Viola macloskeyi*)

• 1 to 3 inches
• early season
• moist meadows, stream edges

This white-flowered species grows in rhizomatous patches from lake level into the subalpine zone. The small leaves are ovate to round with cordate to reniform bases. The lower three petals have noticeable purple veins. At high elevations, Maclosky's violets are often found blooming in July or August, forming scattered flecks of white amidst the late-season, grassy terrain.

Mountain violet (*Viola purpurea* ssp. *integrifolia*)

• 1 to 5 inches
• early season
• dry slopes, open forest

This widespread species is particularly common in the north Basin, surviving in dry habitats into the alpine zone with the help of an especially long taproot. The dark green, deeply veined basal leaves are ovate to scallop-shaped and entire to slightly dentate or wavy-margined. The stem leaves are more lanceolate and entire-margined. The two upper flower petals are purplish on the outside. The closely related **pine violet** (*Viola pinetorum* ssp. *pinetorum*) is found in dry, upper montane habitat in the south Basin. It has small white hairs that give its herbage a lighter, gray-green appearance and narrower, longer, ovate to linear basal and stem leaves, which are borne on a more elongated, branching stem. The uncommon **Shelton's violet** (*V. sheltonii*) occurs in deep gravelly soils on slopes

Beckwith's violet

Maclosky's violet

Stream violet

or in forests up to 8,000 feet. This easy-to-recognize, yellow-flowered species has ternate leaves with many dissected, linear to obovate leaflets.

MISTLETOE FAMILY (VISCACEAE)

This small family of perennial herbs and shrubs is parasitic on woody plants, obtaining inorganic nutrients and water through spe-

Mountain violet

cialized organs that grow directly into the tissues of the host. The brittle stems are swollen at the

Western dwarf mistletoe

native genera are present in Tahoe. Typical of this family, the best way to identify Tahoe mistletoes is to identify their host plants.

Western dwarf mistletoe (*Arceuthobium campylopodum*)

• **2 to 6 inches**
• **mid-season**
• **habitats of host species**

This species is relatively common throughout the upper montane zone, where it grows on pines and firs, typically white fir. **American dwarf mistletoe (*Arceuthobium americanum*)** grows almost exclusively on lodgepole pine at lower elevations. **Juniper mistletoe (*Phoradendron juniperinum*)** occurs on Sierra juniper from low elevations into the subalpine zone. **Incense-cedar mistletoe (*P. libocedri*)** is rare in the Basin, but has been found in white fir forest on the lake's west shore, parasitizing incense-cedar.

joints and have minuscule (sometimes scale-like) opposite leaves and unisexual, radial flowers usually borne on separate *dioecious* plants. The flowers are green, lack petals, and bloom on densely clustered, axillary spikes or cymes. The ovary is inferior. The gelatinous, berry-like fruits are highly prized by birds, that act as dispersal agents by depositing undigested and often sticky coated seeds onto potential host plants. Both of California's

Glossary

See Drawings

achene: Dry, one-seeded fruit from a one-chambered ovary; fruit of sunflower family

alpine: Above timberline; generally encountered in Tahoe between 9,000 to 10,500 feet.

alternate: Arranged singly or in an alternating fashion along an axis (typically a stem).*

angiosperm: Plant that bears true flowers; composed of monocots and dicots.

annual: Plant that completes life cycle in one growing season; non-woody root growth.

anther: Pollen-forming portions of a stamen.*

ascending: Curving or angling upwards from base; less than vertical.*

asymmetric: Irregular in shape; not divisible into identical or mirror-image halves.

banner: The largest, uppermost petal of many members of the pea family, Fabaceae.*

basal: Found at base of plant or plant part; said of leaves clustered at stem base.*

biennial: Completing life cycle in two growing seasons; flowering in second year.

bisexual: Flowers with both fertile stamens and fertile pistils.*

bract: Small leaf- or scale-like structure found below sepals on inflorescence or on cones.*

bulb: Short underground stem and attached fleshy leaves or leaf bases.*

calyx: Collective term for sepals; outermost whorl of flower parts; generally green.*

catkin: Spike of unisexual flowers, typically hanging, often without petals and sepals.*

cauline: Borne on a stem; not basal; typically refers to stem leaves.

circumboreal: Found around the world at northern latitudes.

clone: Genetically identical individual resulting from asexual reproduction.

compound: Composed of two or more parts; e.g., *compound leaf* composed of leaflets.*

cordate: Heart-shaped; often said of a leaf base with rounded lobes.*

corolla: Collective term for petals; whorl of flower parts immediately inside or above calyx.*

cotyledon: Modified leaf present in the seed, often functioning as food storage.

deciduous: Falling off at end of growing season; also said of plants that are seasonally leafless.

decumbent: mostly prostrate, but with tips curving up.

dentate: Having margins with sharp, relatively coarse teeth pointing outward, not tipward.*

dicot: Flowering plant with two cotyledons, flower parts in fours or fives, non-parallel venation.

dioecious: Male and female plants separate.

diploid: Having two sets of chromosomes, maternal and paternal; 2n.

discoid: Flower head in sunflower family composed entirely of disk flowers.*

disk flower: Bisexual, radial flowers in sunflower flower heads.*

dissected: Irregularly, sharply and

deeply cut, but not compound.

elliptic: In the shape of an ellipse.*

entire: Having margins that are continuous and smooth (i.e., without teeth, lobes).*

evergreen: Never lacking green leaves; leaves do not fall all together.

exceeding: Surpassing another structure tipward.

exserted: Protruded out of surrounding structures.

fibrous: Said of slender root system composed of many similar roots.

filament: Anther-stalk; the often thread-like portion of a stamen.*

forked: Branching into two parts of about equal size (compare *stellate*).

free: Not fused to other parts, distinct, separate (compare *fused*).

fringed: Having ragged or finely cut margins.

fruit: A ripened ovary, or collection of ripened ovaries and associated structures.

fused: United or joined together, e.g., petals fused into a corolla tube (compare *free*).

gland: Small structure that exudes a generally sticky substance, on surface or tips of hairs.

glaucous: Covered with whitish or bluish, waxy or powdery film that can be rubbed off.

gymnosperm: Plants with seeds not borne in ovaries but in cones or naked on branches.

haploid: Having one set of chromosomes (maternal or paternal); n.

hastate: Arrowhead-shaped, with two perpendicular basal lobes.*

head: Dense, often spheric inflorescence of flowers.

herb: Plant with non-woody aboveground parts; includes annuals, biennials, and perennials.

herbage: The non-woody, above-ground plant parts, especially leaves, young stems.

hypanthium: Fused lower portions of sepals, petals, and stamens.*

indusium: In many ferns, outgrowth of leaf surface or margin covering the sporangia.*

inferior ovary: Ovary appearing below free flower parts due to fusion of the hypanthium.*

inflorescence: An entire cluster of flowers; generally excluding full-sized leaves.

intergrade: To merge gradually through a more or less continuous series of intermediates.

internode: Segment of axis between two successive attachment points for appendage.*

involucre: Group of bracts occurring below a flower or inflorescence.*

keel: Ridge along long axis of structure; two lowermost, fused petals of pea family members.*

lanceolate: Narrowly elongate, widest in the basal half and tapered to an acute tip.*

leaflet: One leaf-like unit of a compound leaf.*

ligulate: Bisexual flower in ligulate sunflower head (compare *ray flower*).*

linear: Elongate, with nearly parallel sides.*

lip: Upper or lower of two parts in an unequally divided calyx or corolla.*

lobe: Expansion or bulge along margin; free tips of otherwise fused sepals or petals.

monocot: Flowering plant with one cotyledon, flower parts in threes, parallel leaf venation.

monoecious: Male and female unisexual flowers on same plant.

native: Occurring in an area not as a consequence of human activity; indigenous.

naturalized: Non-native, but reproducing without human fostering.

nutlet: Small dry one-seeded fruit encased in a hard shell, usually more than one per flower.

ob: prefix indicating inversion of shape.*

oblong: Longer than wide, with nearly parallel sides, rounded corners.*

open: Uncongested, diffuse (compare dense).

opposite: Located directly across from; two structures per node (compare alternate).*

ovary: Wider portion of pistil, bearing ovule, normally developing into a fruit.*

ovate: Egg-shaped in two dimensions, widest below middle, as of a leaf.*

ovoid: Egg-shaped in three dimensions, widest below middle, as of a fruit.

ovule: Structure containing an egg; a seed prior to fertilization.

palmate: Radiating from a common point; said of veins, lobes, or leaflets of a leaf.*

panicle: Branched inflorescence in which basal or lateral flowers open first along axis.*

pappus: Aggregate of awns or bristles rising above inferior ovary in sunflower family.*

pedicel: Stalk of an individual flower or fruit.*

peduncle: Stalk of an entire inflorescence.*

perennial: Living more than two growing seasons; typically non-woody above ground.

perianth: Calyx and corolla collectively, whether or not they are distinguishable.*

persistent: Not falling off, remaining attached.

petiole: Leaf stalk, connecting leaf blade to stem.*

phyllary: In sunflower family, a bract of the involucre, below the flower head.*

pinnate: Two structures on opposite sides of an axis; said of leaves (compare palmate).*

pistil: Female reproductive structure of a flower; includes ovary, style, and stigma.*

pollination: Placement of pollen on structure through which fertilization occurs.

polyploid: Having three or more sets of chromosomes (compare diploid, haploid).

prostrate: Lying flat on the ground.

protandrous: Bisexual flowers in which pollen is released prior to stigma receptivity.

raceme: Unbranched inflorescence of pedicled flowers that open from bottom to top.*

radial: Divisible into mirror-image halves in three or more ways.

radiate: A sunflower flower head composed of central disk flowers and marginal ray flowers.*

ray: A primary, radiating axis, as a primary branch in a compound umbel.*

ray flower: Long outer margin flower in radiate sunflower head.*

receptacle: Structure to which individual flower parts, flowers, or flower heads are attached.*

recurved: Gradually curved downward or backward.

reduced: Gradually smaller, often narrower, less lobed, etc.

reflexed: Abruptly bent or curved downward or backward.

reniform: Kidney-shaped, as of a leaf.*

rhizome: Underground, horizontal stem; may have leaves, leaf scars, scales, buds, etc.*

root: Underground structure of plant,

generally growing into ground from the base of stem.

rosette: Radiating cluster of leaves generally at or near ground level.*

rotate: Saucer shaped; said of fused corolla with short or absent tube and spreading lobes.*

salverform: Having a slender tube and abruptly spreading corolla lobes.*

seed: A fertilized ovule, earliest product of sexual reproduction in plants.

sepal: Individual member of calyx, usually green and below petal on flower.*

serrate: Margins with sharp teeth generally pointing tipward, not outward.*

sessile: Stalkless; without a petiole, peduncle, or pedicel.

sheath: Structure that surrounds or partly surrounds another structure.

shrub: Woody plant of relatively short maximum height, much branched from base.

simple: Composed of a single part; undivided, unbranched.

sorus (sori): Distinct cluster of sporangia in many ferns.*

spike: Unbranched inflorescence of stalkless flowers, opening from bottom to top.*

sporangium (sporangia): Spore-producing organ on ferns and other non-seed plants.*

spore: Reproductive unit of non-seed plants; haploid cell from diploid parent plant.

spreading: More or less perpendicular to axis of attachment; typically horizontal.

stamen: Male flower part, composed of stalk-like filament and pollen-producing anther.*

staminate: Having fertile stamens but sterile or missing pistils.

staminode: Sterile stamen, often modified in appearance.

stellate: Star-like; hair with three or more branches radiating from common point.

stigma: End of pistil, on which pollen is deposited, elevated above the ovary on a style.*

stipule: Appendage at base of petiole, generally paired, typically leaf- or scale-like.

stolon: Thin runner stem lying flat on the ground, forming roots and erect stems or shoots.*

stomate (stomata): Minute pore on a leaf or stem through which gases pass by diffusion.

style: Stalk-like portion that connects ovary to stigma in many pistils.*

subalpine: Just below timberline; between montane and alpine.

subshrub: Plant with lower stems woody; upper stems not woody; dying back seasonally.

superior ovary: Ovary appearing above free flower parts, which rise from ovary base.*

tapered: Gradually narrower or smaller at base or tip.*

taproot: Main, tapered root that grows straight down into soil, with lateral branches.

taxon (taxa): Group of organisms at any rank, i.e., species, family, etc.

tendril: Slender, coiling structure by which a climbing plant becomes attached to its support.

ternate: Lobed or compounded into three parts, like a clover leaf.

throat: In flowers with fused sepals or petals, the expanded, fused portion above the tube.

tooth: A small, pointed projection of a margin.

tube: In flowers with fused sepals and petals, the cylindric, fused portion at the base.

tubercle: Small, wart-like projection.

umbel: Inflorescence in which three to many pedicels radiate from a common point.*

unisexual: Having flowers in which either stamens or pistils, but not both, are fertile.

vascular: Pertaining to plant veins; plants with veins.

weed: Generally non-native plant, often adapted to disturbed areas.

whorl: Groups of three or more leaves, flower parts etc. of same kind at one node.*

wing: Thin, flat extension of surface or margin; two lateral petals of pea family flowers.

Native Tahoe Vascular Plant Families Not Covered in this Book:

Callitrichaceae (water starwort)

Ceratophyllaceae (hornwort)

Cyperaxceae (sedge)

Elatinaceae (water-wort)

Euphorbiaceae (spurge)

Gramineae (grass)

Haloragaceae (water-milfoil)

Hydrocharitaceae (waterweed)

Isoetaceae (quillwort)**

Juncaceae (rush)

Lemnaceae (duckweed)

Limnanthaceae (meadowfoam)

Marsileaceae (clover fern)**

Moraceae (duckweed)

Potamogetonaceae (pondweed)

Selaginellaceae (spike-moss)**

Sparganiaceae (bur-reed)

**Non-seed

PLANT PARTS AND STRUCTURES

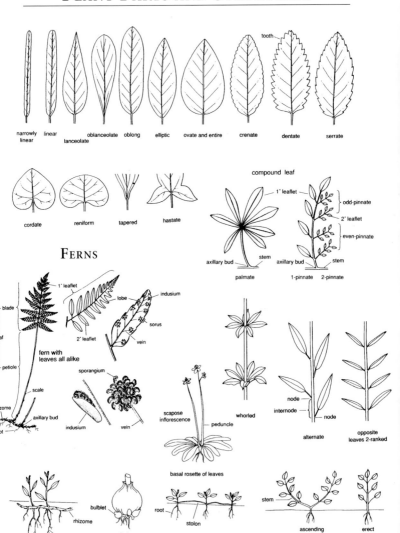

narrowly   linear   oblanceolate   oblong   elliptic   ovate and entire   crenate   dentate   serrate
linear   lanceolate

tooth

cordate   reniform   tapered   hastate

compound leaf

1° leaflet
2° leaflet
odd-pinnate
even-pinnate

axillary bud   stem   axillary bud   stem
palmate   1-pinnate   2-pinnate

FERNS

1° leaflet   lobe   indusium
blade   sorus
leaf   2° leaflet   vein
fern with
leaves all alike
petiole

scale
rhizome   axillary bud
root

sporangium
indusium   vein

scapose
inflorescence   peduncle

whorled

node
internode   node

alternate   opposite
leaves 2-ranked

basal rosette of leaves

bulblet
rhizome   root   stolon   stem
bulb   ascending   erect

FLOWER PARTS AND STRUCTURES

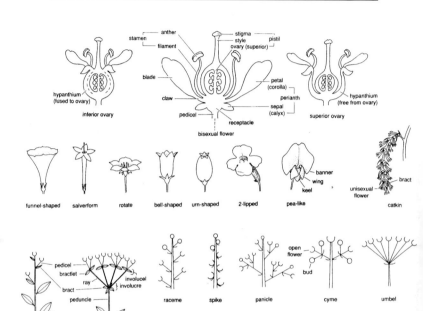

SUNFLOWER PARTS

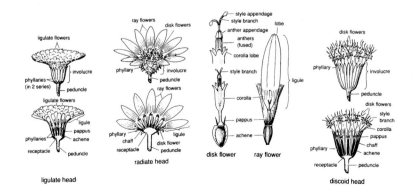

Western buttercups bloom in early season amidst the aspens near South Shore.

Selected References

Bailey, L. 1963. *How Plants Get Their Names*, Dover Press, New York, NY.

Bakker, E. 1984. *An Island Called California.* University of California Press, Berkeley, CA.

Blackwell, L. 1997. *Wildflowers of the Tahoe Sierra*, Lone Pine Publishing, Redmond, WA.

Carville, J. 1989. *Lingering in Tahoe's Wild Gardens.* Mountain Gypsy Press, Chicago Park, CA.

Chabot, B., and W. Billings. 1972. Origins and ecology of the Sierran alpine flora and vegetation. *Ecol. Monog.* 42: 163-199.

Coffey, T. 1993. *The History and Folklore of North American Wildflowers,* Houghton Mifflin, New York, NY.

Fauver, T. 1992. *Wildflower Walking in Lakes Basin of the Northern Sierra,* Fauver & Steinbach, Orinda, CA.

Hickman, J., ed. 1993. *The Jepson Manual of Higher Plants of California.*, University of California Press, Berkeley, CA.

Hill, M. 1975. *Geology of the Sierra Nevada.* California Natural History Guides, no. 37. University of California Press, Berkeley, CA.

Horn, E. 1976. *Wildflowers 3: The Sierra Nevada.* Touchstone Press, Beaverton, OR.

Hutchinson, J., and G. Stebbins. 1986. *A Flora of the Wrights Lake Area.* Judy L. Hutchinson, Pollack Pines, CA.

Niehaus, T., and C. Ripper. 1976. *A Field Guide to Pacific States Wildflowers,* Houghton Mifflin, New York, NY.

Ornduff, R. 1974. *Introduction to California Plant Life,* California Natural History Guides, no. 35. University of California Press, Berkeley, CA.

Potter, B. 1983 *A Flora of the Desolation Wilderness, El Dorado County,* *California*, Masters Thesis, Humboldt State University, Arcata, CA.

Raven, P., and D. Axelrod, 1978. *Origins and Relationships of the California Flora*, University of California Publications in Botany, no. 72. University of California Press, Berkeley, CA.

Schaffer, J., 1987. *The Tahoe Sierra: A Natural History Guide to 106 Hikes in the Northern Sierra*, Wilderness Press, Berkeley, CA.

Schoenherr, A., 1992. *A Natural History of California*, California Natural History Guides, no. 56, University of California Press, Berkeley, CA.

Sharsmith, C. 1940. *A Contribution to the history of the alpine flora of the Sierra Nevada*, Ph.D. Dissertation. Univ. Calif. Berkeley, CA.

Smiley, F. 1915. The alpine and subalpine vegetation of the Lake Tahoe region. *Botanical Gazette* 59(4):265-286.

Smith, G. 1973. A flora of the Tahoe Basin and neighboring areas and supplement. *The Washman Journal of Biology* 31(1):1-231; 41(1), (2): 1-46.

Storer, T., and R. Usinger. 1963. *Sierra Nevada Natural History.* University of California Press, Berkeley, CA.

Taylor, D. 1976. *Ecology of the timberline vegetation at Carson Pass, Alpine County, California*, Ph.D. Dissertation. Univ. Calif. Davis, CA.

Taylor, D. 1976. Disjunction of Great Basin Plants in the northern Sierra Nevada. *Madroño* 23: 301-310.

Weeden, N. 1981. *Sierra Nevada Flora.* Wilderness Press, Berkeley, CA.

Whitney, S. 1979. *A Sierra Club Naturalist's Guide to the Sierra Nevada.* Sierra Club Books, San Francisco, CA.

A view of Hawkins Peak from Grass Lake at Luther Pass.

INDEX